Estimate! Calculate! Evaluate!
Calculator Activities for the Middle Grades

Marjorie W. Bloom
Grace K. Galton

Cover design by Arthur Celedonia

Illustrations by Joshua Berger

Copyright©1990 by
Cuisenaire Company of America, Inc.
12 Church Street, Box D, New Rochelle, New York 10802

CONTENTS

INTRODUCTION

The National Council of Teachers of Mathematics has taken the position that all students should be given the opportunity to use calculators in mathematics classes. The authors fully concur with this point of view and this book has been written to provide constructive ways in which the calculator can be used by middle grade students to facilitate the learning of mathematical concepts.

Calculators are:

- intriguing and appealing to students
- inexpensive and easily obtainable

Calculators permit students to:

- focus on problem solving rather than on tedious computation
- practice and validate their estimation skills
- discover patterns in mathematics more readily
- review and check their recall of number facts

Calculators are useful tools to build the self-image of those students whose ability to solve problems out-distances their ability to compute.

A strong mathematics program should include the teaching of number theory, estimation, statistical techniques, geometry and measurement as well as the basic operations of arithmetic. By freeing students from excessive tiresome computing, calculators make it possible for teachers of middle grade mathematics to include a greater variety of mathematical topics in their programs.

ESTIMATE! CALCULATE! EVALUATE! © 1990 CUISENAIRE COMPANY OF AMERICA, INC.

HOW TO USE THIS BOOK

The worksheets in this book are intended to support a middle grade mathematics program. Each sheet contains a series of problems which develop one or more mathematical concepts. It is the concept or concepts which is the heart of the worksheet. The arithmetical computations necessary to arrive at accurate answers are important, of course, but of far greater importance is the grasp of the concepts themselves. The role of the electronic calculator here is a significant one. It handles the "nitty-gritty" while the student can concentrate on the problem situation and devise processes necessary to find accurate, thoughtful, reasonable solutions.

On the front of each worksheet the problem, game or activity is explained to the student. Generally enough space is given so that the student may work on the page itself. Occasionally a separate record or answer sheet may be provided.

On the back of each worksheet is information for teachers. This includes the objective of the sheet, some comments about the task and some suggestions for teaching, solutions if they are appropriate, and challenge problems. Many teacher pages include suggestions for discussion about alternative problem-solving strategies, about reasonableness of answers, and/or about real world applications. We believe such discussions provide another dimension in which students can be helped to grasp new concepts. Also, while some students learn easily working alone, others benefit from small group work and partnership collaboration. Most of the activities here should be very useful to the teacher whose classroom management strategies include cooperative learning.

We would like to recommend that teachers encourage estimation in most of these activities before allowing the calculator to be used. We have tried to suggest some specific places where estimation would be valuable, but space limitations prevent us from including reminders on all teacher pages.

We hope that you will find these pages useful as teaching materials, as support materials, or as challenges. After you have used them for a while, you will probably develop your own extension materials. We would be delighted to hear from you about the ways in which you are finding the calculator of value in your classroom.

TO THE TEACHER

The activities in this book are written for use with a general-purpose calculator with memory (M+, M−, CM/RM/MRC/R-CM). The calculator should have all the operations keys (+, −, ×, ÷), a separate equals key (=), and a decimal point key (.). We would also recommend that the calculator have a percent key (%), and a square root key (√‾‾‾). For those students whose calculators do not have these special keys, we will suggest alternate ways to solve the problems. A change sign key (+/−) may also be useful to those teachers who would like to use calculators in their work with integers.

Since different brands of calculators vary in appearance and in some functions, we will discuss only those features which are significant for this workbook. If your school is going to purchase calculators, we do recommend that you check to be sure your calculators have certain features we consider important:

- a clear, distinct 8-digit display
- keys which are large enough for small, uncoordinated fingers and which are easy to press
- floating decimal point
- easily observable error message
- memory

During the preparation of this workbook we have used inexpensive general-purpose calculators. We found that they fulfilled all our requirements and that they are relatively easy to use.

A scientific calculator has many additional features such as rounding, scientific notation, and use of conventional rules of order – but we do not think these features are needed by intermediate level students.

For the student there are introductory pages describing or explaining the use of various keys, the functions to be used, and the error message. Use of some keys such as percent and square root will be explained on the teachers' pages following the worksheets where their use is required.

Most fraction activities included here use decimal fractions rather than common fractions. It is our belief that the general-purpose calculator is primarily of value in converting common fractions to decimal fractions and is less helpful in operations with common fractions.

ESTIMATE! CALCULATE! EVALUATE! © 1990 CUISENAIRE COMPANY OF AMERICA, INC.

TO THE STUDENT

LEARN ABOUT YOUR CALCULATOR

Although calculators made by different companies may look different, most of the general-purpose calculators work in similar ways. They all have electronic chips which are the "brains" of the calculators. They all have screens where numbers and/or letters appear. The numbers and/or letters form the display.

Some calculators are powered by batteries which have to be replaced after they have been used for many hours. Battery-operated calculators usually have some type of on/off button. In order to make the batteries last, the user must remember to turn off the calculator after each use. Other calculators are powered by solar cells. These will continue to work as long as enough light – daylight or artificial light – reaches the solar-cell window.

ABOUT THOSE KEYS

If your calculator has these keys:

You should press:		If you want to:
AC or CA	all clear	completely clear the calculator, including the contents of memory
C	clear	clear the contents of the calculator, but not the memory
CE	clear entry	get rid of the last numerical value you entered
ON/C	on and/or clear	turn on the calculator and/or clear the calculator except for memory
M+	add to memory	add a value to the number stored in memory
M−	subtract from memory	take away a value from a number stored in memory
MR	read memory	read what is stored in memory
MRC or R-CM	read memory and clear memory	read what is stored in memory (press once) and clear out what was stored in memory (press twice)
$\sqrt{}$	square root	find the square root of a number
%	percent	change a value to a percent
+/−	change sign	change the sign of a number on the screen to its opposite

TO THE STUDENT

CONSTANT FUNCTIONS

Your calculators may have special constant functions.

Addition and Subtraction

To find out whether your calculator has addition and subtraction constant functions, try the following keys:

$$7 + 3 = = =$$

Did you get 10, 13, 16? If you did, your calculator does have an addition constant function. Notice it keeps adding 3 (the second addend) each time you press $=$. In other words, a constant value of 3 is added in again and again.

If your calculator has the addition constant, it should also have the subtraction constant. Try:

$$7 - 5 = = = =$$

If you don't have the addition and subtraction constant functions, what should you do? Any problems where constants are suggested can be solved by repeating $+$ and the constant value. Try:

$$7 + 3 + 3 + \text{ or } 7 + 3 + 3 + 3 =$$

Notice that 10 appears after you press the second $+$ and that each new sum appears every time you press $+$. You can use this same method to do repeated subtraction.

Multiplication and Division

Now test to see if your calculator has a constant function for multiplication. Try:

$$3 \times 5 = = =$$

Did you get 15, 45, 135? Then you have the multiplication constant function. Notice that it keeps multiplying by the first factor, 3, each time you press $=$. 3 is the constant factor.

Try this division problem:

$$75 \div 5 = =$$

Notice that the constant divisor is 5.

Again, if you do not have the multiplication constant, you can also achieve the same result by pressing $\times$ and the number. Try:

$$3 \times 5 \times 3 \times 3 \times \text{ or } 3 \times 5 \times 3 \times 3 =$$

ERROR MESSAGES

What happens if the number you want to put in is too big for the calculator? Try:

$$42{,}568 \times 3{,}812 =$$

On your screen there will be the letter E along with some of the digits in the answer. This message tells you that the answer is too large for the display and that it has overflowed the calculator. This only happens when the number is larger than 99,999,999.

On the other hand, you will not get an error message if your answer contains more than 8 digits to the right of the decimal point. Try:

$$2 \div 3 =$$

Note that the answer is given as 0.6666666. Some calculators round the 7th decimal place – so the display reads 0.6666667. Most calculators, however, just drop any digits after the 7th decimal place.

If you press the following sequence of keys,

$$9 \div 0 =$$

you will also get an error message. In the mathematics we use, division by zero is not allowed. Problems such as $9 \div 0$ have no meaning.

ESTIMATE! CALCULATE! EVALUATE! © 1990 CUISENAIRE COMPANY OF AMERICA, INC.

WHERE DO I BELONG?

Next to each row of boxes is a set of digits. Place the digits in the boxes to produce the given product.

1. ☐ ☐ ☐ (2, 3, 4)
× ☐ ☐ ☐ (4, 5, 6)
─────────
1 5 6, 7 3 5

2. ☐ ☐ ☐ ☐ (4, 5, 8, 9)
×   ☐ ☐ (5, 6)
─────────
4 7 3, 7 0 4

3. ☐ ☐ ☐ ☐ (2, 4, 6, 7)
×   ☐ ☐ ☐ (2, 4, 5)
─────────
2, 9 2 5, 3 4 4

4. ☐ ☐ ☐ ☐ (1, 2, 7, 8)
×   ☐ ☐ ☐ (2, 3, 6)
─────────
7 9 1, 6 9 4

5. ☐ ☐ ☐ ☐ (1, 4, 5, 8)
×   ☐ ☐ ☐ (2, 4, 5)
─────────
2, 3 2 6, 8 9 6

6. ☐ ☐ ☐ ☐ (4, 6, 7, 8)
× ☐ ☐ ☐ ☐ (5, 7, 8, 9)
─────────
6 1, 0 6 9, 2 8 0

WHERE DO I BELONG?

OBJECTIVE: The student will use estimation to facilitate placing factors to produce a given product.

COMMENTS: Students should first be given the opportunity to experiment with one or two problems on their own. Then the class should share any methods they use to simplify the process. Some ideas which they present may seem strange or different but may be helpful to other students in thinking their way through the task.

For those who have difficulty, the teacher may want to suggest looking first at the ones digit in the product. What pairs of factors will produce products with the same ones digit? In problem #1, only 3 and 5 will produce a 5 in the ones place.

Another clue may be to look at the digits in the largest places in the number. In problem #1, for example, which pairs of remaining digits will give a product close to but not larger than 15? Either 2×4 or 2×6 would work, but 2×6 is closer.

Encourage estimation and educated guessing as strategies to solve these problems. It is probably a good idea to suggest to students that they record the factors as they try them so that they will know which combinations they have already tried.

SOLUTIONS:

1. $\times$ 243 645	**2.** $\times$ 8459 56	**3.** $\times$ 6472 452
4. $\times$ 2187 362	**5.** $\times$ 5148 452	**6.** $\times$ 7648 7985

ADDITIONAL ACTIVITY:

This activity could be set up as a game for two or more students. Each student could make up five problems to give to an opponent. (The teacher might want to set limits on the number of digits which may be used.) The opponent would start with 25 points. For each incorrect trial, the player must deduct one point from his/her score. The player who ends up with the highest score is the winner.

MYSTERY BOXES

The first of the two puzzles below is a multiplication puzzle. The second is a division puzzle.

In the multiplication puzzle, the top horizontal row and the left vertical column are the factors. In the division puzzle, the top horizontal row is the dividend and the left vertical column is the divisor.

Fill in the empty boxes in the grids below.

Puzzle 1

×		475	86	
12				1,068
			4,644	
97	54,514			
		86,450		

Puzzle 2

÷		80,028	12,996	
228	2,052			
57				1,368
		468		
			38	

MYSTERY BOXES

OBJECTIVE: The student will investigate the relationship of multiplication and division by finding factors and multiples.

COMMENTS: This is a wonderful opportunity to reinforce the concept of inverse processes and review vocabulary terms such as factor, multiple, product, dividend, and divisor. Students should discuss strategies to use to find the missing boxes.

SOLUTIONS:

Puzzle 1

✕	562	475	86	89
12	6,744	5,700	1,032	1,068
54	30,348	25,650	4,644	4,806
97	54,514	46,075	8,342	8,633
182	102,284	86,450	15,652	16,198

Puzzle 2

÷	467,856	80,028	12,996	77,976
228	2,052	351	57	342
57	8,208	1,404	228	1,368
171	2,736	468	76	456
342	1,368	234	38	228

ADDITIONAL ACTIVITY:

Have students make up charts of their own for their classmates to do.

ZERO IN ON THE TARGET

Challenge a classmate to play this game with you. Each of you will need a calculator and a ZERO IN ON THE TARGET score sheet (page 86).

45	103	199	346
154	487	416	522
219	57	228	89

The game consists of six rounds.

In Round 1 each of you chooses, writes down and *adds* five numbers from the list above. The object is to get as close as you can to the target number (400 in the sample given below). You can be above or below the target. Once you have selected your numbers, you may not make any changes. After you obtain your total, find and write down the difference between your total and the target. This difference is your score. You may use the same number as many times as you would like.

Rounds 2 and 3 are played the same way you played Round 1.

In Rounds 4-6 you may *add or subtract*, as you choose, to get to the target.

The winner is the person who has the lowest point total at the end of the 6 rounds.

Sample of each type of round:

SCORE SHEET

ROUND #	TARGET	MY NUMBER	MY TOTAL	OPPONENT TOTAL	DIFFERENCE BETWEEN EACH TOTAL & TARGET	
					MINE	OPPONENT
1	400	45 +103 +154 + 89 + 57	448	351	48	49
4	1200	487 +522 +228 -199 +154	1192	1184	8	16
				Totals after 2 sample rounds:	56	65

As you can see, you would be leading at this stage of the game, but not by much.

ZERO IN ON THE TARGET

OBJECTIVE: The student will estimate sums and differences of whole numbers to get as close as possible to a target number.

COMMENTS: This game is written for two students but actually three or four could play at the same time.

After the students have reviewed the directions, teachers might start with an example such as a target number of 500. Several students might be asked to offer estimates which would be written on the chalkboard or overhead projector. After the totals have been obtained and the best selections determined, a discussion of the strategies would be helpful to all. It should be stressed to the students that they must write down all the numbers they are going to use, before they try them. Also, once they have entered a number in the calculator, they may not change it.

The winner, of course, is the student with the *lowest* total. For this game, only the absolute value of the difference between the player's score and the target number is used. For example, let us assume that the target number is 500. Player 1's total is 650. Player 2's total is 350. Both players are 150 points from the target. Therefore they have the same score, 150 points.

For the subtraction rounds, teachers should use their own discretion as to whether they will allow the students to work with negative partial totals.

See Appendix (page 86) for the score sheet.

SOLUTIONS: Not applicable

ADDITIONAL ACTIVITY:

There are many possible extensions of this game. The teacher might challenge the students to find the "best" combinations, the combinations which will actually come closest to the target numbers.

A different set of target numbers or a different set of twelve working numbers can be used.

The authors used this game with great success as an entire class activity, using a separate sheet of about 50 working numbers in a 4×12 or 5×10 grid form. In this version of the game no number could be used more than once. Teachers and/or students might want to make up such a sheet for use in their own classrooms.

MULTIPLICATION DETECTIVE

Use your estimation and pattern-finding skills to help you solve each of these problems.

1. Find the multiplication problem which produces each product, using the directions given.

> **Example:** Use six 1's to form two factors whose product is 12,221.
>
> **Solution:** $1,111 \times 11 = 12,221$

   a. Use four 1's and one 0 to form two factors whose product is 1,111.

   _____

   b. Use five 1's and two 0's to form two factors whose product is 111,111.

   _____

2. Using each set of digits below, form two factors which will produce the greatest product. (Note: There is a pattern, but it is not as obvious as it may seem.)

> **Example:** If you use the digits 1, 2, and 3, the pair of factors which will produce the greatest product is 21 and 3.

   a. 1, 2, 3, 4, 5 (Hint: The answer is larger than 22,405.)

   _____ $\times$ _____ = _____

   b. 1, 3, 5, 7, 9

   _____ $\times$ _____ = _____

   c. 0, 2, 4, 6, 8

   _____ $\times$ _____ = _____

   d. 1, 2, 3, 4, 5, 6, 7 (Challenge!)

   _____ $\times$ _____ = _____

3. Use only single-digit *prime* numbers to fill in the boxes in this problem. The digits may be used more than once.

$$\square \ \ \square \ \ \square \ \times \ \square \ \ \square \ = \ \square \ \ \square \ \ \square \ \ \square \ \ \square$$

MULTIPLICATION DETECTIVE

OBJECTIVE: The student will use place value, placeholder concepts, estimation and pattern-discovery skills to find the missing factors in multiplication problems.

COMMENTS: The first problem is quite easy once students think about where the 0 *cannot* be placed. Solutions to problem #2 are less obvious. In #2a, for example, while it is likely that students will select a three-digit factor and a two-digit factor, the optimum placement of the digits is less clear. Most students try to place the 5 in the three-digit factor – 5xx × xx. The hint for the first part should convince them that this will not provide the largest product. The teacher might suggest that the student check the answer to the first one before proceeding to the remainder of this section.

The most difficult problem on this sheet is problem #3 where each box must be replaced by a single-digit prime number (2, 3, 5, and/or 7). To solve this, students should be encouraged to keep systematic records of their trials, so that they will not keep repeating the same combinations. Also, by keeping records they should be able to analyze the results and determine what digits cannot be in particular places. For example, they should quickly discover that 2 cannot be in the ones place in either factor, because the ones place in the answer would not contain a prime number.

It is worthwhile to have students verbalize their method or methods of finding solutions. These problems lend themselves very well to classroom discussion. Some teachers might prefer to ask students to explain in writing their strategies for solving one or more problems.

SOLUTIONS:

1. a. $101 \times 11 = 1,111$

   b. $1,001 \times 111 = 111,111$ or $10,101 \times 11 = 111,111$

2. a. $431 \times 52 = 22,412$

   b. $751 \times 93 = 69,843$

   c. $640 \times 82 = 52,480$

   d. $6,531 \times 742 = 4,846,002$

  For any different digits, A, B, C, D, and E, where A is the smallest and E is the largest, the pattern is DCA × EB. For any different digits, A, B, C, D, E, F, G, the pattern is FECA × GDB.

3. $775 \times 33 = 25,575$

ADDITIONAL ACTIVITY:

Fill in each box in this problem with a single-digit *prime* number:

☐ ☐ ☐ ☐ × ☐ ☐ ☐ = ☐ ☐ ☐ ☐ ☐ ☐

Solution: $7,735 \times 333 = 2,575,755$

SPACIOUS STATES

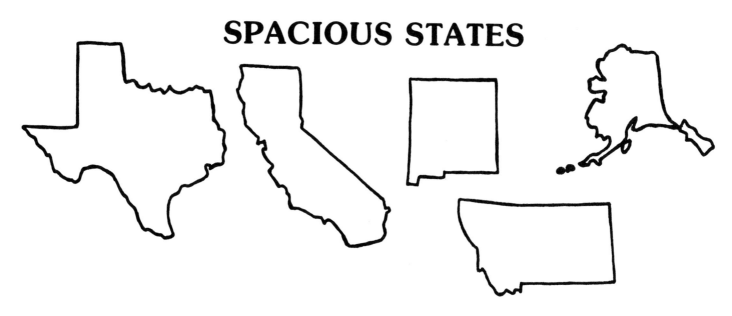

Use the clues given below to determine the five states with the largest land area. List the states in order of decreasing size and write the land area of each one.

State	Land Area
1.	
2.	
3.	
4.	
5.	

Clues:

1. The second largest state is almost one-half the size of the largest.

2. Montana has a land size of 147,046 square miles.

3. The fifth largest state has 37,113 fewer square miles than the third largest state.

4. Alaska has a land area which is 432,298 square miles larger than the third largest state.

5. The third largest state is in the same position in the list as its starting letter is in the alphabet.

6. Texas has 145,214 square miles more than the fifth largest state.

7. California is only 11,660 square miles bigger than Montana.

8. New Mexico is also one of the five largest states.

SPACIOUS STATES

OBJECTIVE: The student will analyze a set of clues in order to solve a logic problem.

COMMENTS: This worksheet uses the logic problem format where students must read each clue carefully and compare and evaluate information. Most students will have had some experience with this type of problem. However, for those who do not know where to begin, the teacher might point out clue #5. The third largest state and the third letter of the alphabet should lead the student to discover that the state of California fits into this position. To find out how big California is, the student must then use the information given in clues #2 and #7. In this manner, the pupil should be able to complete the chart.

SOLUTIONS*:

1. Alaska   591,004

2. Texas   266,807

3. California   158,706

4. Montana   147,046

5. New Mexico   121,593

ADDITIONAL ACTIVITY:

Students might try to write their own clues for the following set of information* about the five states with the *smallest* areas.

Rhode Island   1,212 square miles

Delaware   2,045 square miles

Connecticut   5,018 square miles

Hawaii   6,471 square miles

New Jersey   7,787 square miles

*Source: *The World Almanac and Book of Facts, 1989*

PURCHASING POWER

Mr. Clark's class posted this price list for school supplies.

MR. CLARK'S CLASS
PRICE LIST
Posters......... 85¢ ea. 3-Ringed
Pencils......... 10¢ ea. Notebook.. $2.75
Index Cards...35¢ ea. Typewriter
Pens.......... 25¢ ea. Paper...... $2.25
Pencil   Erasers...... 15¢ ea.
  Sharpener...$1.65 Labels........ 50¢ ea.
                Stapler...... $3.50

Use the price list to solve the following problems:

1. Betsy bought a stapler, three pencils, an eraser, and two labels. How much money did she need?

2. How much money would you need to buy one of each item?

3. Jenny bought five pencils, three labels, two pens, a poster, and a package of typewriter paper. Sally purchased three packages of index cards, five posters, and an eraser. Who spent more money? How much more?

4. What is the greatest number of items you can buy for $5.00 if you can buy no more than two of each item?

5. Randy bought four items for $3.95. What did he buy?

6. Ann brought $5.00 to school. She bought an equal number of pens and pencils and received a dime in change. How many of each did she buy?

7. The class needed to know what was left in the store at the end of the year. They found the most they had of any selection was three and that they were totally out of four of the selections. The total value of the items remaining in the store amounted to $16.65. Determine what was left in the store.

Is there more than one solution? How do you know?

PURCHASING POWER

OBJECTIVE: The student will locate and use information from a chart. The student will also learn to simplify a problem using estimation.

COMMENTS: Below are some questions which could be asked as an introduction to this worksheet.

The first three problems provide practice in using the memory key to store part of an answer. When more than one of a particular item is needed, the total price can be determined by multiplication and the result can then be stored in memory.

Example: How much would a package of index cards and three notebooks cost?

Solution: To solve it using the memory, the students should press the following keys:

.35 $\boxed{M+}$ 2.75 $\boxed{\times}$ 3 $\boxed{M+}$ $\boxed{R-CM}$

Answer: $8.60

Some students may be able to solve the last four problems by trial and error, but a follow-up discussion about how different students solved these problems should make it clear that the better one estimates, the easier it will be to solve the problem. Students should also have the opportunity to compare answers. They should discover that for problems #5 and #7 there are several possible correct responses.

Example: What items could you buy if you were asked to spend $5.00 exactly?

Solution: A three-ringed notebook and typewriter paper *or* a stapler and 3 labels

SOLUTIONS:

1. $4.95

2. $12.35

3. Jenny, $.15

4. 12 (2 pencils, 2 erasers, 2 pens, 2 packages index cards, 2 labels, and 2 posters)

5. Answers will vary. Here are two possible solutions:
   a. 1 notebook, 1 label, and 2 packages index cards
   b. 1 stapler, 1 pen, and 2 pencils

6. 14 pens, 14 pencils

7. Answers will vary. Here are three possible solutions:
   a. 3 notebooks, 3 packages typewriter paper, 2 pencils, 1 pen, 1 poster, 1 package index cards
   b. 2 packages typewriter paper, 2 notebooks, 2 posters, 3 labels, 2 pencil sharpeners, 1 eraser
   c. 3 posters, 2 notebooks, 2 packages typewriter paper, 1 stapler, 1 label, 1 pencil

ADDITIONAL ACTIVITY:

Have the students use a newspaper advertisement to make up problems to challenge one another.

LEFTOVERS

$15 \div 4 = 3\frac{3}{4}$ I group   I group   I group   ¾ group

If you use the calculator to solve the problem pictured above, your screen will show 3.75.

The two results, 3¾ and 3.75, are the same since ¾ = 0.75.

The calculator does not give you the exact remainder, but there are several ways to find it using the calculator. Can you find one way to get the exact remainder of 3? (Hint: What is ¾ or .75 of 4?)

Find the exact remainder for each of these division problems.

1. $2196 \div 40$ **2.** $795 \div 32$ **3.** $4169 \div 57$

Remainder _____   Remainder _____   Remainder _____

Use the four numbers given below to solve problems #4, #5, and #6. For these problems you can only divide larger numbers by smaller numbers (no zero quotients).

1290   634   296   80

4. Which two numbers will form a division problem which leaves the largest remainder?

_____

What is the remainder? _____

5. Which two numbers will form a division problem which leaves the smallest remainder?

_____

What is the remainder? _____

6. What is the remainder when you divide the largest five-digit number by the smallest three-digit odd number?

_____

After you have solved some of the problems on this worksheet, compare your method with the methods devised by your classmates. Did you all do these the same way? Is there an easiest way?

_____

_____

LEFTOVERS

OBJECTIVE: The student will understand the relationship of the decimal fraction to the remainder in a division problem.

COMMENTS: It may be helpful to use the following simple division problem to show that there is more than one way to solve a division problem and to determine the exact remainder in a division problem, using the calculator. You may want to present the problem first, ask students to try it on the calculator, and seek comments on the answer the calculator gives. Can the students come up with any suggestions as to how they can find the exact remainder?

$$22 \div 8 = ?$$

On the calculator the answer reads 2.75. There are a number of ways to determine that 6 is the exact remainder. Probably the first way is the best way for showing students the relationship between the decimal fraction and the exact remainder. This method will not given an exact remainder when the decimal fraction is a repeating decimal, because of the rounding limitations of the simple, general purpose calculator. However, the student should be able to determine the remainder by rounding up the whole number. (Note: This rounding error does not occur if you use a scientific calculator.)

1. $22 \div 8 = 2.75$.   There are two 8's in 22 and .75 of an 8.
$$.75 \times 8 = 6$$
$$22 \div 8 = 2 \text{ R } 6$$

2. $22 \div 8 = 2.75$.   There are two 8's in 22 and something left over.
$$2 \times 8 = 16$$
$$22 - 16 = 6$$
$$22 \div 8 = 2 \text{ R } 6$$

3. $22 \div 8 = 2.75$ Students may use calculator memory to solve these problems.

22 $\boxed{\text{M}+}$

2×8 $\boxed{\text{M}-}$ (or $2 \times 8 =$ $\boxed{\text{M}-}$, depending on the calculator)

$\boxed{\text{RCM}}$

4. Still another way is by repeated subtraction. Place 22 in the calculator and count the number of 8's you subtract. Stop when you reach a number less than 8 (6). This is the remainder. The counted total is the number of 8's in 22. We do not recommend this method as it is too difficult to keep track in division of large numbers.

SOLUTIONS:

1. 36 (54.9)

2. 27 (24.84375)

3. 8 (73.14035)

4. $1290 \div 296$
106 (4.3581081)

5. $1290 \div 80$
10 (16.125)

6. 9 (990.0891)
$99999 \div 101$

ADDITIONAL ACTIVITY:

If students have difficulty with the large numbers, teachers may prefer to give them a set of problems using smaller numbers.

STRANGE SPORTS

Every year the Blooperville Gazette publishes the statistics for the best individual school sports records.

This year there were some strange "bugs" in the typesetting machine. Use your calculator to fill in the blanks left by the printer. Answers should be rounded as indicated.

Flashy Football Team: Quarterback Passing Records

Name	Number of Completed Passes	Total Yards Passed	Average Number of Yards per Completed Pass (to nearest hundredth)
Pete Penalty	124	529	_ _ _
Larry Lateral	116	487	_ _ _
Sam Stumbletoes	_ _ _	472	4.03

Booming Baseball Team: Best Batting Averages

Name	Number of Times at Bat	Number of Hits	Batting Average (to nearest thousandth)
Billy Butterfingers	_ _ _	18	.383
Bobby Bobble	49	18	_ _ _
Tom Tripper	48	_ _ _	.313

Bouncy Basketball Team: Best Point Scorers

Name	Number of Games Played	Total Points Scored	Average Number of Points per Game (to nearest tenth)
Frank Freethrow	19	_ _ _	17.4
Dan Dribbler	21	343	_ _ _
Ron Ringer	_ _ _	334	15.9

STRANGE SPORTS

OBJECTIVE: The student will compute averages for sports in their common decimal form.

COMMENTS: Both the meaning of an average and the method of computation should be reviewed before attempting this page.

In football, the average number of yards per completed pass is obtained by dividing the total yards passed by the number of completed passes.

In baseball, a batter's average is obtained by dividing the number of hits he makes by the number times he is at bat. While it is somewhat difficult to see the batting average as a true average, it can be looked at in the following way: A turn at bat is either successful (1) or not (0). Six attempts at bat might be recorded as:

$$\frac{1 + 0 + 0 + 1 + 0 + 0}{6} \text{ or } \frac{2}{6} = .333$$

Batting averages are always given to nearest thousandth.

In basketball, the average number of points per game is obtained by dividing the total points scored by the number of games played.

SOLUTIONS:

124	529	4.27
116	487	4.20
117	472	4.03
47	18	.383
49	18	.367
48	15	.313
19	331	17.4
21	343	16.3
21	334	15.9

ADDITIONAL ACTIVITY:

Students can pose problems by using the sports statistics from the newspaper. Have them calculate their own averages from some of their own experiences.

SEARCHING FOR THE PATTERN

Where do we go from here? "Here" is the last number in each set of numbers below. Your calculator can help you in your search for the next two numbers in each sequence. Sometimes only one operation will do it. Sometimes you will need to use more than one.

Example: 18   9   4.5   2.25   1.125   _____   _____

Solution: The pattern is division by 2. Once you know the pattern, the rest is easy. Use your calculator to find the next two numbers. Did you get 0.5625 and 0.28125?

Now try these.

1. 2   2.4   2.88   3.456   _____   _____

2. 1.2   2.7   5.7   7.2   10.2   _____   _____

3. 22.8   19.2   23.5   19.9   24.2   _____   _____

4. 6.3   11   15.7   20.4   25.1   _____   _____

5. 63.5   55.8   48.1   40.4   32.7   _____   _____

6. 132   99   74.25   55.6875   41.765625   _____   _____

7. 5   15   21.75   65.25   72   216   _____   _____

8. 7.2   13.2   26.4   23.4   29.4   58.8   55.8   _____   _____   _____

SEARCHING FOR THE PATTERN

OBJECTIVE: The student should be able to discover the pattern which determines each sequence and then find more members of the sequence.

COMMENTS: Here are some examples which could be used on the chalkboard or overhead projector to help students improve their skill at finding patterns. Some of the techniques they could use are 1) trial and error or 2) testing the first two numbers by adding, subtracting, multiplying or dividing one into the other.

As students discover the pattern, suggest that they show their solutions by giving the next numbers of the sequence, rather than by revealing the operation or operations. In this way, the process will still be "a secret" and other students will have a chance to make the discovery for themselves.

Example:

a.   59,049   19,683   6,561   2,187   _____   _____

b.   19   20.5   22   23.5   25   _____   _____

c.   1200   480   192   76.8   30.72   _____   _____

Solution:

Divide by 3;   729   243

Add 1.5;   26.5   28

Multiply by .4 or multiply by 2 and divide by 5; 12.288   4.9152

SOLUTIONS:

Note: Answers given are those shown on the calculator display.

1.	4.1472; 4.97664	$\times$ 1.2
2.	11.7; 14.7	+ 1.5, + 3
3.	20.6; 24.9	− 3.6, + 4.3
4.	29.8; 34.5	+ 4.7
5.	25; 17.3	− 7.7
6.	31.324218; 23.493163	($\div$4, $\times$ 3) or $\times$ 0.75
7.	222.75; 668.25	$\times$ 3, + 6.75
8.	61.8; 123.6; 120.6	+ 6, $\times$ 2, − 3

ADDITIONAL ACTIVITY:

1.   16   22   34   58   106   _____   _____
Solution: 202; 394; rule is $2n - 10$

2.   50   80   125   192.5   293.75   _____   _____
Solution: 445.625; 673.4375; rule is $1.5n + 5$

LOOKING IN THE MIRROR

See if you can discover the answer pattern. The title of this page and the numbers in the first two boxes should provide a clue.

Fill in each box and place the correct number on the blank line.

1. $9.2 + 7.065 + 38.46 =$ $\boxed{54.725}$

   _____ $+ 108.93 - 152.935 =$ $\boxed{527.45}$

2. $679.43 - 52.728 + 254.068 =$ $\boxed{}$

   $116.037 -$ _____ $+ 57.29 =$ $\boxed{}$

3. $54.9 - 2.682 - 0.497 =$ $\boxed{}$

   $417.485 - ($ _____ $+ 15.963) + 23.048 =$ $\boxed{}$

4. $27.62 + (748.2 - 576.29) =$ $\boxed{}$

   _____ $- (35.49 + 23.143) =$ $\boxed{}$

5. $6291.073 + 476.6 + 0.923 -$ _____ $=$ $\boxed{5802.316}$

   _____ $- (42.4 - 3.072 + 27.23) =$ $\boxed{}$

6. $2.8 \times 4.6 \div$ _____ $=$ $\boxed{8.05}$

   $3.175 \div$ _____ $\times 0.8 =$ $\boxed{}$

7. $369 \times 0.24 \div 40 =$ $\boxed{}$

   $72.135 \times 3.6 \div (90 \times$ _____ $) =$ $\boxed{}$

LOOKING IN THE MIRROR

OBJECTIVE: The student will try to discover a mathematical pattern and will use all four operations with decimal fractions.

COMMENTS: Encourage children to investigate all possibilities in trying to discover a pattern. Remind them that the title of the activity is a clue. A discussion of what mirrors do to images might be helpful to discover that the answers in each pair of equations are reflections of one another. In the reflected number, the decimal point appears the same number of places from the right as it occurred from the left in the original number. For example, the reflection of 4.12 is 21.4.

Example: $722 + 0.09 - 486.9 + 6170.1 = \boxed{6405.29}$

$54.208 - 0.0003 + \underline{\hspace{2cm}} - 1.43 = \boxed{\underline{\hspace{2cm}}}$

Note the reflection of the decimal point.

Solution: 39.7269; 92.5046

SOLUTIONS:

1. 571.455

2. 880.77
96.239   77.088

3. 51.721
297.42   127.15

4. 199.53
94.624   35.991

5. 966.28
679.7665   613.2085

6. 1.6
0.05   50.8

7. 2.214
0.007   412.2

ADDITIONAL ACTIVITY:

Students should try to make up some of these for other classmates to try. They are not as easy to create as they look.

DARTS, IF YOU DARE

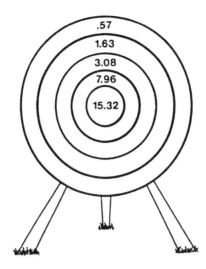

You have 8 darts to work with! Follow the directions for each problem. All the darts you use must hit the target. No misses allowed!

1. What is the largest total you could get, using all 8 darts?

_____

What is the smallest total you could get, using all 8 darts?

_____

2. Game 1 is a 3-dart game. Show the closest you can get to a total of 5 without going over it.

_____

3. Game 2 is a 5-dart game. Show the closest you can get to a total of 25 without going over it.

_____

4. Use the fewest darts you need to get a total of exactly 42. What numbers did you hit?

_____

5. What is the largest whole number you can hit using 5 darts? Which numbers did you use?

_____

What is the largest whole number you can hit using 7 darts? Which numbers did you use?

_____

6. What is the smallest whole number you can hit using however many darts you need (up to a maximum of 8 darts)? Which numbers did you use?

_____

7. The winner of this game is the one who shoots a score closest to 12, but not necessarily lower than 12. How many darts do you need?

_____

What numbers must you hit? What is your total?

_____

DARTS, IF YOU DARE

OBJECTIVE: The student will solve problems requiring addition of decimals, using the problem-solving strategy of educated guessing.

COMMENTS: Many of the problems on this worksheet are quite difficult and require that the student try many combinations before deciding on a final answer. It is suggested that the student write down each combination as he/she tries it in order to make the most efficient use of time.

It is important to point out the rules. There are at most 8 darts which can be used. Some problems, however, specify fewer darts. A number can be hit over and over again.

This would be a good worksheet for students to work in pairs or in threes. They might each try combinations and then compare answers. When they have completed a few of the problems, it might be advisable for them to check with the teacher to determine whether they are on target, close to the answer, or still far away from it.

SOLUTIONS:

The authors believe that they have found the best answers, but they would appreciate hearing from you if your students find an answer which is even more appropriate.

1. 122.56
   4.56

2. $1.63 + 1.63 + 1.63 = 4.89$

3. $15.32 + 7.96 + .57 + .57 + .57 = 24.99$

4. 7 darts
   $5(7.96) + 1.63 + .57 = 42$

5. 57; $3(15.32) + 7.96 + 3.08 = 57$
   95; $6(15.32) + 3.08 = 95$

6. 19; $2(7.96) + 3.08 = 19$

7. 6 darts; $3(3.08) + 1.63 + 2(.57) = 12.01$

ADDITIONAL ACTIVITY:

The students can create any number of new challenges by changing the number of darts allowed, or by putting a different set of numbers on the target.

ESTIMATE AND CALCULATE
A Game for 2 or 3 Players

Each player estimates the answer to the same problem and writes it down. The problem is done on the calculator and a player's score is the difference between his/her estimate and the exact answer. Each player enters his/her score for each round. One game consists of six rounds. The winner is the person with the *lowest* score at the end of the game. You will need one score sheet for all players. (See page 87).

GAME 1

1. 4.22 × 5.52 = _____

2. 3.29 × 15 = _____

3. 23.7 × 4.2 = _____

4. 9.8 × 0.35 = _____

5. 12 × 1.9412 = _____

6. 6.82 × 3.7 = _____

GAME 2

1. 15.08 × 15.78 = _____

2. 0.67 × 9716 = _____

3. 0.273 × 5.7 = _____

4. 5297 × 0.006 = _____

5. 2.11 × 0.86 = _____

6. 8.7 × 0.903 = _____

Now try picking a factor to reach the given answer!

GAME 3

1. 124 × _____ = 102.92

2. 0.17 × _____ = 3.944

3. 4.93 × _____ = 387.991

4. 0.058 × _____ = 54.926

5. 3.249 × _____ = 591.318

6. 4023 × _____ = 2977.02

GAME 4

1. 0.4 × _____ = 1138.8

2. 12.6 × _____ = 167.454

3. 100.34 × _____ = 5762.5262

4. 61672 × _____ = 15603.016

5. 2.7 × _____ = 0.02592

6. 0.125 × _____ = 2653.75

Now choose both factors!

GAME 5

1. _____ × _____ = 224.1438

2. _____ × _____ = 657.305

3. _____ × _____ = 0.0228

4. _____ × _____ = 15.232

5. _____ × _____ = 1.26324

6. _____ × _____ = 11.2744

GAME 6

1. _____ × _____ = 4.5216

2. _____ × _____ = 3186.196

3. _____ × _____ = 263.055

4. _____ × _____ = 113.078

5. _____ × _____ = 1.184

6. _____ × _____ = 805.93521

ESTIMATE AND CALCULATE

OBJECTIVE: The student will estimate products of decimal fractions by rounding.

COMMENTS: Have the children estimate some sample problems and ask one or two of the children to explain how he/she arrived at the answer. It is a good idea to review rules of rounding. Emphasize reasonableness of estimate. Differences should be expressed as positive values (absolute value) for this game.

> **Example:** A guess of 9.5 for 3.4 × 2.7 = _____ would be scored as .32 since 9.5 − (3.4 × 2.7) = .32. Note that a guess of 2.5 and 3.928 would also score as .32 since (2.5 × 3.928) − 9.5 = .32.

Note: Use score sheet on page 87.

SOLUTIONS:

GAME 1

1. 23.2944
2. 49.35
3. 99.54
4. 3.43
5. 23.2944
6. 25.234

GAME 2

1. 237.9624
2. 6509.72
3. 1.5561
4. 31.782
5. 1.8146
6. 7.8561

GAME 3

1. 0.83
2. 23.2
3. 78.7
4. 947
5. 182
6. 0.74

GAME 4

1. 2847
2. 13.29
3. 57.43
4. 0.253
5. 0.0096
6. 21230

For games 5 and 6, answers will vary.

ADDITIONAL ACTIVITY:

Teachers and/or students can make up additional problems using the operation of division. This version of the game is quite difficult to play because of the frequent occurrence of repeating decimal quotients. You and your students may agree to round answers to the nearest thousandth before calculating individual scores.

HI....   33

Choose only from the list of numbers below. Fill in the boxes in each sentence to find the largest possible answer for each problem. Each number may be used only once in an individual problem.

2.8   3.42   1.5   0.9

1. [___] + [___] − [___] = _____

2. [___] × [___] + [___] = _____

3. [___] − [___] × [___] = _____

4. [___] ÷ [___] + [___] = _____

Check: Add up the values thus far. The sum should be 25.066.

5. [___] ÷ [___] − [___] = _____

6. [___] × [___] − [___] = _____

7. [___] ÷ [___] × [___] = _____

8. [___] + [___] ÷ [___] = _____

Check: Add up all eight numbers. If your answers are correct, the sum should be 53.282.

HI....

OBJECTIVE: The student will show understanding of arithmetic processes using decimal fractions.

COMMENTS: Remind students about rules of order.* It is helpful to point out what happens when they multiply or divide by numbers less than one.

*From left to right, do 1) parentheses, 2) exponents, 3) multiplication and division as they occur and 4) addition and subtraction as they occur. A more detailed explanation of rules of order appears in the comments on the teacher sheet for **Parentheses** (page 61).

SOLUTIONS:

These are the best solutions the authors have obtained thus far, but they would appreciate knowing if your students do better.

1. $2.8 + 3.42 - 0.9 = 5.32$

2. $3.42 \times 2.8 + 1.5 = 11.076$

3. $3.42 - (1.5 \times 0.9) = 2.07$

4. $3.42 \div 0.9 + 2.8 = 6.6$

5. $3.42 \div 0.9 - 1.5 = 2.3$

6. $3.42 \times 2.8 - 0.9 = 8.676$

7. $3.42 \div 0.9 \times 2.8 = 10.64$

8. $2.8 + (3.42 \div 0.9) = 6.6$

ADDITIONAL ACTIVITY:

Other numbers can be substituted at any time to provide an ongoing activity. This could be used in game format.

....LO

Choose only from the list of numbers below. Fill in the boxes in each sentence to find the smallest possible answer for each problem. Each number may be used only once in an individual problem. Only positive answers are allowed. Round to the nearest thousandth where necessary.

2.8   3.42   1.5   0.9

1. [   ] + [   ] − [   ] = _____

2. [   ] × [   ] + [   ] = _____

3. [   ] − [   ] × [   ] = _____

4. [   ] ÷ [   ] + [   ] = _____

Check: Add up the values thus far. The sum should be 6.539.

5. [   ] ÷ [   ] − [   ] = _____

6. [   ] × [   ] − [   ] = _____

7. [   ] ÷ [   ] × [   ] = _____

8. [   ] + [   ] ÷ [   ] = _____

Check: Add up all eight numbers. If your answers are correct, the sum should be 8.872.

....LO

OBJECTIVE: The student will show understanding of arithmetic processes using decimal fractions.

COMMENTS: Remind students about rules of order.* Note that negatives are not allowed in this activity.

*From left to right, do 1) parentheses, 2) exponents, 3) multiplication and division as they occur and 4) addition and subtraction as they occur. A more detailed explanation of rules of order appears in the comments on the teacher sheet for **Parentheses** (page 61).

SOLUTIONS:

These are the best solutions the authors have obtained thus far, but they would appreciate knowing if your students do better.

1. $2.8 + 0.9 - 3.42 = .28$

2. $2.8 \times 0.9 + 1.5 = 4.02$

3. $3.42 - (2.8 \times 0.9) = 0.9$

4. $1.5 \div 3.42 + 0.9 = 1.339$

5. $3.42 \div 2.8 - 0.9 = .321$

6. $3.42 \times 0.9 - 2.8 = .278$

7. $1.5 \div 3.42 \times 0.9 = .395$

8. $0.9 + (1.5 \div 3.42) = 1.339$

ADDITIONAL ACTIVITY:

Other numbers can be substituted for these problems. You may also want to change the rules to allow for negative solutions. This activity can be used in game format for either an entire class or a smaller group.

CALLING ALL CABS

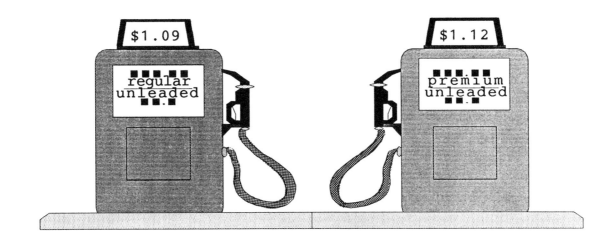

Wally Waldo owns three large cabs. They all use special premium unleaded gasoline. He and his drivers each work six days a week.

The cab Wally drives is a real gas guzzler. Last week it used 14 gallons of gas per day on the average. He drove 105 miles per day.

Jake's cab averaged 122 miles per day and got 20 miles to the gallon last week.

Wendy only drives from the hotel to the airport and back again. The distance to the airport is 13.5 miles. She made 6 round trips a day last week and averaged 21.5 miles to the gallon.

1. What was the total cost for gasoline for all Waldo Taxi Service drivers last week?

_____

Jerry Boyd owns a small fleet of mini-cabs. Two of his cabs use regular unleaded gas and three use premium unleaded gas. He and his drivers also work six days a week.

Jerry's car used $32.70 worth of regular gas last week. He averaged 24.5 miles per gallon.

Sally averaged 126 miles per day last week, using premium gas. She used 6 gallons a day.

Two other mini-cabs, using premium gas, together covered 2,280 miles last week. They each averaged 19 miles per gallon.

Lloyd's cab uses regular gas. His cab was in for repairs last Thursday, but he averaged 126 miles per day for the other five days. His gas cost $7.84 per day on the average.

2. What was Boyd Cab's total gasoline expense for last week?

_____

3. Which taxi company had a lower gasoline cost per mile traveled last week?

_____

4. How much lower? _____

CALLING ALL CABS

OBJECTIVE: The student will solve a variety of multiple-step rate problems.

COMMENTS: To complete this sheet successfully, students must read the problems carefully and keep the information they obtain in an organized fashion. Problems #1 and #2 ask for the cost of gasoline for each company for a six-day week. Note that one cab, because of repairs, only works a five-day week.

The problems on this worksheet are variations of rate problems.

Distance = rate (miles per day) × time (days per workweek) or

Distance = rate (miles per gallon) × number of gallons

Sometimes students will need to find the rate, sometimes the distance. Several steps may be required. Students will also have to decide which information is relevant and necessary to solve a problem. To answer problems #3 and #4, students will have to go back to the original data and total the number of miles all the cabs have driven during the week. They will then be able to determine the rate (cost per mile) and which cab company has a lower gasoline cost. Again, the necessary data is there, but the solution will require some analysis and evaluation of the information by the students.

SOLUTIONS:

1. Expenditures/week for Waldo Taxi Service:

Wally – $94.08   Jake – $41.00 ($40.992)   Wendy – $50.64 ($50.6344...)   Total – $185.72

2. Expenditures/week for Boyd Cab:

Jerry – $32.70   Sally – $40.32   2 mini-cabs – $134.40   Lloyd – $39.20   Total – $246.62

3. Boyd Cab had a lower gasoline cost per mile traveled.

Boyd Cab – 4,401 miles traveled; cost – about $.06/mile
Waldo Taxi Service – 2,334 miles traveled; cost – about $.08/mile

4. About $.02/mile less

BATTERS UP

In 1941, Ted Williams ended the baseball season with a batting average of .406. However, in the last 40 years, the batting champions of both the American and the National League have had averages which ranged between .300 and .400.

Use the clues below to place the top 6 players in order from highest to lowest and determine each player's average.

Player	**Batting Average**
1.	
2.	
3.	
4.	
5.	
6.	

Clues:

1. The difference between the top player on the list and the sixth player is very small. The sixth player's average is 94.3% of the first player's average.

2. The second and third players on the list are tied.

3. There is only a difference of .001 between the fifth and sixth players.

4. In the year when Wade Boggs achieved his highest batting average, he was at bat 653 times and hit safely 240 times.

5. The difference between George Brett's average and Stan Musial's average is .014.

6. Ted Williams' average is .019 greater than Tony Gwyn's.

7. Rod Carew is the third player on the list.

8. Musial's average is .008 higher than Boggs'.

BATTERS UP

OBJECTIVE: The student will solve a logic problem, given a set of clues which use decimal fractions.

COMMENTS: Most students have worked with this type of logic problem. Students should be instructed to read all the clues carefully before starting on the problem. If they do not realize that clue #4 is the logical starting place, they may need some guidance finding it. They will need to use the information given there to calculate Wade Boggs' batting average (.3675344 ≈ .368). From this they can then compute Stan Musial's average (.376). This, of course, still does not make it clear where these players rank, but we can tell that Musial is not in 5th or 6th place from the information given in clue #3. Now it is possible to determine George Brett's average (.390). Thus far, the lowest average belongs to Wade Boggs. It is still not clear that Boggs is at the bottom of the list. However, students might now check to see whether Boggs' average is 94.3% of Brett's average...and, of course, it is. Students should then be able to complete the remainder of the problem.

SOLUTIONS*:

1. George Brett   1980   .390

2. Ted Williams   1957   .388

3. Rod Carew   1977   .388

4. Stan Musial   1948   .376

5. Tony Gwyn   1987   .369

6. Wade Boggs   1985   .368

*Source: ***The World Almanac and Book of Facts, 1989***

UPS AND DOWNS

At least every six months, Nancy checks her investments in the stock market. The stocks she owns and their prices can be seen in the chart below.

Name Of Stock	Symbol	Number Of Shares	Price 12/88	Price 6/89
Black & Decker	BDK	205	23⅛	20¼
Borden	BN	87	59¼	69¾
Coca Cola	KO	68	44⅝	57½
Cray Research	CYR	79	60⅝	53
Delta Airlines	DAL	52	50⅛	67⅝
Eastman Kodak	EK	76	45⅛	50⅞
General Electric	GE	65	44¾	54⅝
IBM	IBM	45	121⅞	109¼
Quaker Oats	OAT	106	53⅜	61⅞

1. What was the total value of Nancy's stock at the end of the year 1988? In June, 1989?

2. What was the amount of the profit or loss since December on Nancy's total investment?

3. Which stock had the greatest rise per share, and how much is Nancy's profit on the stock?

4. Which stock provided the greatest loss per share, and how much is Nancy's loss on the stock?

5. Which stock gave Nancy the largest increase of all her investments and how much is this?

6. Which stock provided the greatest overall loss? How much?

7. If Nancy had sold everything in December and had used the money to buy just one stock, what is the largest profit she could have made by June?

8. What is the most Nancy could have lost?

9. In June, 1989, you inherited $10,000.00. You invested your money in two stocks from the list above. Check your newspaper today to see how you've done. Did you make or lose money? How much?

UPS AND DOWNS

OBJECTIVE: The student will change a common fraction to a decimal fraction before solving a series of problems.

COMMENTS: A discussion about the stock market page and how to gather information on an individual stock is necessary as a preliminary activity for most students.

It should be noted that we have given two answers in some of the problems due to rounding. In problems #1 and #2 the stock broker would round the total cost of each individual stock before totaling the entire list. However, some students may use the calculator memory to obtain their totals. Individual stocks will not be rounded and the answers will be slightly different.

SOLUTIONS:

1. $37,806.14   ($37,806.11)
   $40,725.13

2. $2,918.99 profit   ($2,919.02)

3. Delta Airlines $910.00

4. IBM $568.13

5. Borden $913.50

6. Cray Research $602.38

7. Delta Airlines $13,195.00 ($37,806.11 ÷ 50.125 = 754)

                              (754 × $17.50 = $13,195.00)

8. Cray Research $4,750.38

9. Answers will vary.

ADDITIONAL ACTIVITY:

Ask each student to pretend that he/she has $10,000.00 to invest. He/she is to choose two common stocks from the newspaper using current prices. After a prescribed period of time, he/she can check his/her profit or loss.

FROM THE LUMBER YARD

1. Jackie has just purchased a rectangular piece of wood whose area is 462 square inches. The length and width of this piece are whole numbers. What are all the possible dimensions it could be?

_____

_____

_____

His piece of wood has a perimeter of 106 inches. What are its exact measurements?

_____

What uses might he have for a piece of wood this size?

_____

2. If Jackie bought a rectangular piece of plywood which had an area of 210 square inches and whole number dimensions, what could be the largest and smallest possible perimeters of his purchase?

Largest? _____ Smallest? _____

3. Casey purchased 92 feet of ceiling molding for one room and 76 feet for a second room. If each room was designed to have the largest possible area and whole number dimensions, what are the dimensions of each room?

Room #1 _____

Room #2 _____

4. Casey also bought 24 feet of baseboard for the three walls of her new sliding door closet. The fourth wall is the sliding door. If the dimensions of the closet are whole numbers, what are all the possible sizes, in feet, it could be?

_____

_____

Which of these sizes would give her the most floor space?

_____

Which closet would you prefer to have? Why?

_____

FROM THE LUMBER YARD

OBJECTIVE: The student will solve area and perimeter problems.

COMMENTS: In order to solve these problems, students should review both area and perimeter concepts. In the first problem, they must list all the possible combinations (factors) which will produce an area of 462 square inches. Then they will be able to deduce the correct size of the wood by finding which of their set of dimensions gives the desired perimeter. The question which asks about the uses of the piece of wood should create good discussion. It will indicate how well students can actually visualize a given set of dimensions. Students might suggest such things as shelving, a table top, a gameboard, and backing for a picture.

Some students may not yet be aware of the relationship between the area and perimeter of rectangles. It is hoped that students know that a square is a special form of a rectangle and that they will be able to verbalize the fact that the closer a rectangle with a given perimeter is to a square, the larger will be the area enclosed. Teachers may need to present additional problems to develop the awareness of this relationship.

To solve the fourth problem, where only part of the perimeter is given, the student must evaluate all the possibilities in order to find the largest possible area. Students may be surprised to find that the solution here, when only part of the perimeter is given, is not a square. Again, the answers to the question which asks about the preferred closet size should raise some interesting points for discussion and should clarify whether students can relate a given set of dimensions to a useful area.

SOLUTIONS:

1. 1″ × 462″   2″ × 231″   3″ × 154″
       6″ × 77″   7″ × 66″   11″ × 42″
      14″ × 33″   21″ × 22″

      11″ × 42″

      Answers will vary.

2. 422 inches (210″ × 1″)
       58 inches (14″ × 15″)

3. Room #1: 23′ × 23′ (529 sq. ft.)
       Room #2: 19′ × 19′ (361 sq. ft.)

4. Width × Depth   Width × Depth   Width × Depth   Width × Depth
       22′ × 1′   20′ × 2′   18′ × 3′   16′ × 4′
       14′ × 5′   12′ × 6′   10′ × 7′   8′ × 8′
        6′ × 9′   4′ × 10′   2′ × 11′   1′ × 12′

      12′ × 6′ (72 sq. ft.)

      Answers will vary.

JOIN THE CLUB

Use your best judgment in rounding your answers so they make sense.

1. The neighborhood health club has an exercise pool 20 feet wide by 40 feet long and 5 feet deep. How many gallons of water are needed to fill it to a level about six inches from the top, if there are approximately 7 gallons of water in one cubic foot of water?

2. In order to keep the pool in good condition, the pool maintenance crew needs to add 3 tablets of a chlorine compound for every 10,000 gallons of water. How many chlorine tablets will they use?

3. There is also a track for jogging. Bill does one lap on the track in 50 seconds. If he runs at the rate of 7.2 mph, how long is the track?

4. How many laps on the same track must Mary complete if she wants to jog for 2 miles?

5. Maggie prefers to use the track to walk. She walks at the rate of 3.9 mph. Approximately how long will it take her to walk 2 miles?

6. The health club has 129 members who use the soft drink bar. Each glass holds 7 ounces of liquid and 4 ice cubes. In order to serve each member an average of 3 soft drinks a week, how many gallons of soft drinks must the health club buy?

7. Jenny rides an exercise bicycle. She likes to ride at least 10 miles on this equipment. If she rides at a speed of 14 mph, how many minutes must she ride?

JOIN THE CLUB

OBJECTIVE: The student will solve measurement problems using ratio and proportion.

COMMENTS: Students will need to know equivalent measures, such as ounces in a gallon, minutes in an hour, etc. There are several methods for solving these problems, and in fact the teacher should initiate a discussion to determine how students actually reached their solutions. One method would be the use of ratio and proportion. The following problem might serve as an introduction to this worksheet.

Carrie walked 3.5 miles at the rate of 4.2 miles per hour. How many minutes did it take her to complete this walk?

To solve this by the proportion method, students might write:

$$\frac{3.5 \text{ miles}}{? \text{ minutes}} = \frac{4.2 \text{ miles}}{60 \text{ minutes}} \quad \text{or} \quad 3.5 \text{ miles} : ? \text{ minutes} = 4.2 \text{ miles} : 60 \text{ minutes}$$

Solution: 50 minutes

A second method might be to *develop* a chart.

Distance	Time
4.2 miles*	60 min.*
0.7 miles	10 min.
2.1 miles	30 min.
2.8 miles	40 min.
3.5 miles	50 min.

*Given data

When students calculate the answers to most of these problems, they will find that the solutions involve several decimal places. However, such answers really do not make much sense. For example, the solution to problem #2 is 7.56 tablets. No one, however, would use .56 of a tablet. The student should reason that 7, 7½, or 8 tablets would make much more sense. If the maintenance crew wanted to avoid over-chlorination they might use only 7 tablets. The tablets might be scored so that half a tablet could be used. Or the crew might reason that they wanted to be absolutely safe and use the extra half tablet. Any of these answers would be appropriate and reasonable.

SOLUTIONS:

1. About 25,200 gallons

2. 7, 7½, or 8 tablets (7.56)

3. 0.1 miles or 528 feet

4. 20 laps

5. Between 30 and 31 minutes (30.76923)

6. 22 gallons, to be sure they don't run out (21.164062)

7. About 43 minutes (42.857142)

'ROUND THE WORLD

Round your answers so they make sense.
Use 3.14 for π.

1. If the earth were a perfect sphere, and the distance around the equator is 24,901 miles, about how far is it from any point on the equator to the center of the earth?

2. Jules Verne wrote a book, *Around the World in Eighty Days* about Phileas Fogg, an imaginary character who won a bet from his peers because he traveled the world in record time, 80 days. This trip supposedly took place from October 2 to December 20, 1872. If Fogg traveled a distance of about 23,000 miles, how many miles did he average per hour?

3. In the late 1970's several pilots attempted to set new records for around-the-world travel. One of them made a trip of 22,985 miles in 57 hours, 25 minutes. How many miles per hour did he average?

4. In 1988 Friendship One, a Boeing 747SP, carrying 100 persons and a car, took off from Seattle and circled the globe in just under 36 hours and 55 minutes. Their trip was approximately 23,000 miles. How many miles per hour did they average?

5. If an airplane flew around the earth at the equator at an altitude of 6 miles, how much farther would it travel than car traveling around the earth on the equator (if this could be done)?

6. Jupiter's diameter is approximately 88,000 miles. How many times more than the diameter of the earth at the equator is this? (Use the answer to Problem #1 to help you.)

7. The earth is approximately 92,900,000 miles from the sun. If the light from the sun takes a little more than 8.3 minutes to reach the earth, how fast does light travel in *miles per second*?

'ROUND THE WORLD

OBJECTIVE: The student will solve circumference problems and distance-rate-time problems.

COMMENTS: Problems #1 and #6 require the student to find the radius or the diameter of a circle. In this situation, the circle is the imaginary circle around the earth which is the equator. Students should know that the earth is really not a perfect sphere, that the equator is not a perfect circle, and that certain liberties have been taken in writing these problems.

Problems #2, #3, #4, #5 and #7 require an understanding of the relationship between rate, time and distance: $d = rt$. Here is a problem which might help students review this concept:

One of the fastest U.S. train runs on record was accomplished by a train traveling between Baltimore, Maryland and Wilmington, Delaware. This train traveled a distance of 68.4 miles at a rate of 93.3 miles per hour. How many minutes did this trip take?

Answer: $d = rt$
$68.4 = 93.3 \times t$
$t = .733$ hour or about 44 (43.98) minutes

SOLUTIONS:

1. About 3,965 (3,965.127) miles

2. Just under 12 (11.979) miles per hour

3. A little over 400 (400.319) miles per hour

4. About 623 (623.025) miles per hour

5. About 38 (37.677) miles more

6. About 11 (11.097) times

7. About 186,546 (186,546.18) miles per second

TIE A YELLOW RIBBON

Jackie's hobby is framing rectangular pictures with ribbon. The length and width measurements of each picture are whole numbers.

1. She has a piece of pleated yellow ribbon 30 inches long. If she uses all the ribbon, what are the dimensions of the picture with the largest area she can frame?

_____

2. One of her pictures has an area of 289 square inches. What is the smallest amount of ribbon she might be able to use?

_____

3. She needs special ribbon sold only by the foot for a picture 48″ × 70″. How many feet must she buy?

_____

4. Joan just bought 5 yards of ribbon. She decided to use it to frame three pictures. What is the largest total area (in square inches) of all the pictures she can frame using all of this ribbon?

_____

What are the dimensions of each picture?

_____

Do you think she really frames pictures of these sizes? Why or why not?

_____

Brad wraps packages in the local department store. Boxes are wrapped in two directions as shown in the picture. In addition to a 20-yard roll of yellow ribbon, the store provides three pre-cut lengths of ribbon, 36″, 60″, and 84″. The dimensions of all the boxes are whole numbers.

5. Which pre-cut length of ribbon would Brad need to use for each of these boxes? (Note: Since the store provides bows and tape, you can ignore the amount necessary for a knot and a bow.) The dimensions below are given in the following order: length × width × height. Do you really need this information? Why or why not?

_____

a. a box 8″ × 8″ × 4″ _____

b. a box 6″ × 8″ × 9″ _____

c. a box 10″ × 4″ × 2″ _____

6. What is the box with the largest volume he could wrap using each of the pre-cut ribbons?

a. 36″ _____

b. 60″ _____

c. 84″ _____

7. What is the smallest amount of ribbon he could use on a package with a volume of 32 cubic inches?

_____

TIE A YELLOW RIBBON

OBJECTIVE: The student will solve area, perimeter, and volume problems.

COMMENTS: One purpose of this worksheet is to help students to recognize that, when the perimeter is fixed, the rectangular shape which will have the largest area is a square. However, when the total distance around a box (as measured by ribbon used to tie up the box) is fixed, the largest possible volume of the box is less easily determined.

Note that the directions state that all picture dimensions are whole numbers. This means, in the first problem, if all 30 inches of ribbon are used, the closest shape to a square, with whole number dimensions, is 8″ × 7″. In the third problem, the exact answer is slightly less than 20 feet, but since the ribbon is only sold by the foot the answer should be adjusted accordingly. In the fourth problem the answer which will produce the largest area is probably not very sensible, since a picture one-inch square is rarely desirable.

Students should realize that it is common to wrap the ribbon around the box in two directions, as in the illustration. In working through problem #5 students may discover that the ribbon length for any box is determined by 2l + 2w + 4h. Knowing how to read dimensions is important when one has to determine the amount of ribbon necessary for wrapping.

Problem #6 will help students recognize that for the largest possible volume, V = Ah, the area (A = l × w) should be as large as possible. This is accomplished when two faces of the box are squares. In order to solve the problem, the length and width of the box must be the same, while the height can be different. Students may find it helpful to make a list of the possible combinations. Such a list is started here:

(2 × l) + (2 × w) + (4 × h)	Ribbon length	Volume (l × w × h)
(2 × 1) + (2 × 1) + (4 × 8)	36″	8 cu. in.
(2 × 2) + (2 × 2) + (4 × 7)	36″	28 cu. in.
(2 × 3) + (2 × 3) + (4 × 6)	36″	54 cu. in.

SOLUTIONS:

1. 7 × 8 (56 sq. in)

2. 68 in. (17″ × 17″ picture)

3. 20 feet (19.666666 ft.)

4. 1,851 sq. in.
43″ × 43″; 1″ × 1″; 1″ × 1″

5. Knowing which is the height of the box is essential. The ribbon is used twice over the length of the box, twice over the width of the box, and four times over the height.

a. 60″ ribbon (actual length needed is 48″)
b. 84″ ribbon (actual length needed is 64″); If students were allowed to use more than one piece of ribbon, it would be more economical to use two 36″ pieces on this box.
c. 36″ ribbon (actual length needed is 36″)

6. a. 108 cu. in. (6″ × 6″ × 3″)
b. 500 cu. in. (10″ × 10″ × 5″)
c. 1,372 cu. in. (14″ × 14″ × 7″)

7. 24″ (4″ × 4″ × 2″)

ADDITIONAL ACTIVITY:

Students might suggest solutions for problem #4 where the sizes of the pictures would be more sensible and the total area of the pictures would be as large as possible under the circumstances.

SQUARE AND NOT SO SQUARE

1. The area of this square is 1156 square inches. How long is each side?

_____

2. The area of this circle is 2640.74 square inches. What is its radius? (Use 3.14 for π.)

_____

3. This figure is composed of two congruent semicircles and a rectangle. The area of the rectangle is 1080 square centimeters. The area of the whole figure is 1786.5 square centimeters. What is its perimeter?

_____

4. There are 6 congruent squares in this figure. The area of the whole figure is 20,184 square centimeters. What is its perimeter?

_____

5. In the figure to the right, the area of the dotted section is 100.48 square centimeters. What is the area of the shaded section?

_____

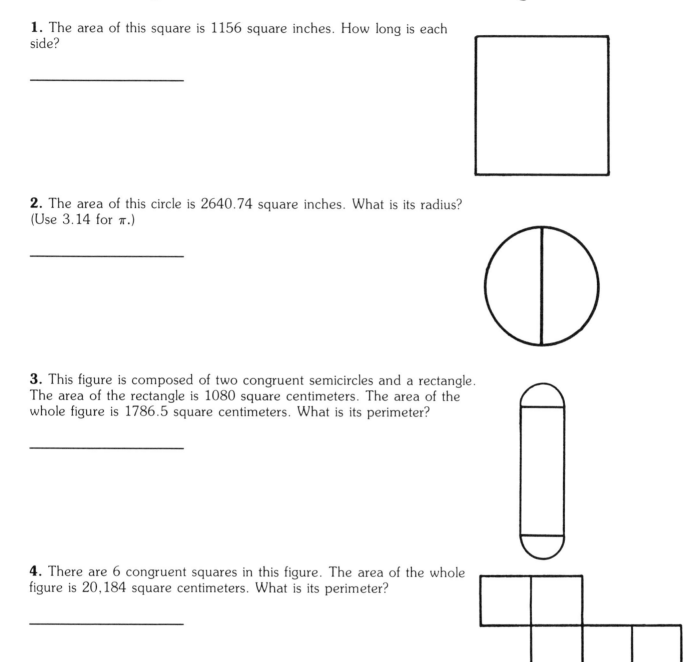

SQUARE AND NOT SO SQUARE

OBJECTIVE: The student will work with areas and perimeters of rectangles and areas and circumferences of circles, using the square root key as needed.

COMMENTS: Students who understand the meaning of *square root* and how to use the square root key will solve the problems on this worksheet most efficiently. It would be helpful to present to them orally a set of numbers such as 81, 144, and 400. It is hoped that some of the students will see the relationship between 81 and 9, 144 and 12, and 400 and 20. It should be clear to them that 81 is the square of 9 and that 9 is the square root of 81.

Most of the problems are self-explanatory. In problem #3, some students may want to tackle each semi-circle separately. Others may realize that the two semi-circles together form a full circle and if they find the circumference of the full circle, they then only have to add in the lengths of two sides of the rectangle.

In problem #5, students could double 100.48 to find the area of the circle. By dividing by 3.14 and taking the square root, they can then find the radius of the circle. The diameter of the circle (2 × radius) is the side of the square. The answer to the problem then can be obtained by subtracting 100.48 from the area of the square.

SOLUTIONS:

1. 34 inches

2. 29 inches

3. 166.2 centimeters

4. 812 centimeters

5. 155.52 square centimeters

ADDITIONAL ACTIVITY:

In the problem below, the area of the circles is 1808.64 square centimeters. What is the area of the rectangle?

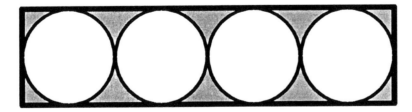

Solution: 2,304 square centimeters

DON'T LET THIS FLOOR YOU!

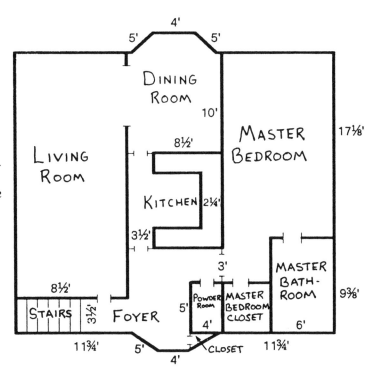

Here is the floor plan of a new house. The builder needs to cover all the floors. He is planning to use carpeting for the living room, dining room, and master bedroom (including the closet). The foyer, the powder room, and the master bathroom will be covered with ceramic tile. In the kitchen the floor surface will be vinyl tile. Your challenge is to find the measurements and amounts of floor covering he needs. All floor coverings are sold in whole units. The builder cannot buy a fractional amount.

1. What is the minimum number of square yards of carpeting needed for each of these rooms?

Living room _____

Dining room _____

Master bedroom _____
(including closet)

2. What is the minimum number of square feet of ceramic tile needed for each of these rooms?

Foyer _____
(including closet)

Powder room _____

Master bath _____

3. What is the minimum number of square feet of vinyl tile needed for the kitchen?

_____

4. What will be the builder's cost for materials at the following prices, including a 5% sales tax?

Dining and living room
carpets @ $18.75/sq. yd.   _____

Master bedroom
carpet @ $16.25/sq. yd.   _____

Foyer ceramic tile @ $ 2.85/sq. ft. _____

Powder room and master bath
ceramic tile @ $ 2.05/sq. ft.   _____

Kitchen vinyl tile @ $ 3.09/sq. ft. _____

Subtotal _____

5% Sales Tax _____

Total _____

DON'T LET THIS FLOOR YOU!

OBJECTIVE: The student will interpret a diagram (floor plan) and determine the dimensions and the areas of different shaped rooms.

COMMENTS: Students should be aware that all the angles in the diagram are right angles, with the exception of the foyer and the dining room. Most of the rooms can be subdivided in such a way that rectangles are formed whose area can be readily determined. In order to work on area, however, they must first determine the linear measurements of some walls which are not labeled. To do this, the students will need to find equal opposite sides whose length and width are given at least in part.

To find the area of the foyer and the bay window area in the dining room, students should know the Pythagorean Theorem. The length of the hypotenuse (5) and the length of one side (4) of the right triangle are given. The area of the right triangle can be found after students use the Pythagorean Theorem to find the third side of the right triangle. Students should be reminded to use care in carrying out their conversions from square feet to square yards. Many forget that there are 9 square feet in a square yard.

It should also be pointed out that, for purposes of this activity, all the flooring materials may be purchased only in full square feet or square yards, not in fractional parts. Students may be tempted to round as they usually do in many rounding situations. However, to be sure that there is sufficient floor covering, it is necessary to "round up." Of course, no account has been taken here of the fact that there is always some waste. Perhaps some of your students will make this observation (see Additional Activity below). Similarly, it should be remembered that the tax is rounded up.

SOLUTIONS:

1. Living room   31 sq. yd.   (270.25 sq. ft.)
    Dining room   17 sq. yd.   (148 sq. ft.)
    Master bedroom   29 sq. yd.   (255.125 sq. ft.)
    (including closet)

2. Foyer (including closet)   112 (111.375) sq. ft.
    Powder room   20 sq. ft.
    Master bathroom   57 (56.25) sq. ft.

3. Kitchen   55 (54.75)sq. ft.

4. Dining room and living room carpet $900.00
    Master bedroom carpet   471.25
    Foyer ceramic tile   319.20
    Powder room and master bathroom tile   157.85
    Kitchen vinyl tile   169.95

                 Subtotal $2018.25
       5% Sales Tax   100.92
                   Total $2119.17

ADDITIONAL ACTIVITY:

In order to account for the waste, students might be asked to add an extra amount (anywhere from 5% to 10%) to each quantity of floor covering and then to recalculate the cost.

54

TV FAVORITES

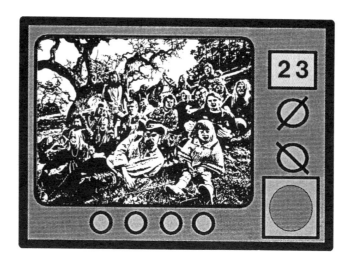

Nielsen Media Research rates the popularity of TV shows. The questions below are based on the Nielsen Report for November, 1987.* Answers should be expressed to the nearest tenth where appropriate.

1. Approximately one out of every five people watching *The Bill Cosby Show* in November 1987 was a teenager. If 4,550,000 teenagers watched the show, how many people watched?

2. Almost five out of every eight people watching *Growing Pains* were under eighteen years of age. If 38,167,584 people watched *Growing Pains*, how many people under eighteen watched the show?

3. For every seven men who watched *A Different World*, there were twelve women who watched. If 33,086,120 adults watch *A Different World*, how many of them were men?

4. 7,181,240 out of 34,360,000 children watched *Alf*. For every sixteen children who watched television, about how many watched *Alf*?

5. *The Bill Cosby Show* was the favorite TV show among women. 24,629,200 of them saw it during the test period. If 91,900,000 women viewed TV, what percent of these women watched *The Bill Cosby Show*?

6. 98% of U.S. homes own TV sets. There are 90,270,000 homes altogether. How many homes do not have TV sets?

7. 85,030,000 TV homes own at least one color TV set. What percent of all U.S. homes own a color TV?

*Source: ***The World Almanac and Book of Facts, 1989***

TV FAVORITES

OBJECTIVE: The student will apply the concept of proportion in word problems.

COMMENTS: Review the concept of proportion presenting both of the following forms:

$$\frac{2}{3} = \frac{8}{12} \quad \text{and} \quad 2:3 = 8:12$$

One way to solve these problems is by writing proportions. For those students who use the proportion form, we would suggest that these problems be solved as equivalent fractions. Some of the more advanced students could attempt solving for the unknown by cross multiplication. It might be a nice idea to have both methods used for checking purposes. Some students may prefer to solve these problems in still other ways. For example, in problem #1, students may realize that 4,550,000 represents ⅕ of the viewers. A discussion of their methods would help the class to recognize that there are generally a variety of methods to reach a solution, not just one right way. Estimating first will help students know if their calculator answers make sense.

SOLUTIONS:

1. $\dfrac{1}{5} = \dfrac{4,550,000}{x}$

x = 22,750,000

2. $\dfrac{5}{8} = \dfrac{x}{38,167,584}$

x = 23,854,740

3. $\dfrac{7}{19} = \dfrac{x}{33,086,120}$

x = 12,189,622

4. $\dfrac{7,181,240}{34,360,000} = \dfrac{x}{16}$

x = 3 (3.34)

5. $\dfrac{24,629,200}{91,900,000} = \dfrac{x}{100}$

x = 26.8%

6. $\dfrac{98}{100} = \dfrac{x}{90,270,000}$

x = 88,464,600

90,270,000 − 88,464,600 = 1,805,400

7. $\dfrac{85,030,000}{90,270,000} = \dfrac{x}{100}$

x = 94.2%

MOTHER'S DAY SPECIALS

It was the Saturday before Mother's Day and Big Deal Appliance Store was having a very special sale for all the last minute shoppers. The sales tax is 5%.

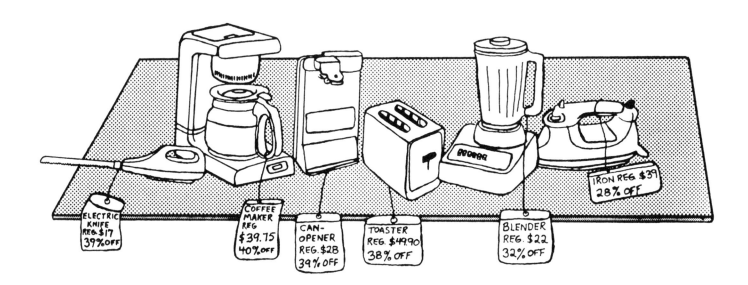

1. Jared had $15. Could he buy the blender? Why or why not?

_____

2. Elsie wanted to buy her mother a new coffee maker. How much money did she need?

_____

3. Mary Alice had $30. What item or combination of items could she buy? Which purchase would leave her with $1.17 in change?

_____

4. Marcus wanted to get his mother a new iron. He worked for the last three weeks cutting grass on Saturday morning. He was paid $9.85 each week. Did he have enough to buy the iron? Why or why not?

_____

5. The twins, Mal and Sal, bought a gift together. They discovered that there was a $10.92 savings on each of two different sale items. They bought the less expensive item. What did they give their mother?

_____

6. a. The Big Deal sold 3 irons, 4 toasters, 2 blenders, 5 coffee makers, 2 electric knives, and 4 can openers on this Saturday. What were the total receipts, including tax, for these appliances on that day?

_____

b. If the store had decided to sell all its items at 36% off (where did this figure come from?) how would its receipts, including tax, compare?

_____

MOTHER'S DAY SPECIALS

OBJECTIVE: The student will solve percent problems.

COMMENTS: The calculator is an excellent tool for finding discounts and for adding on tax. While there is a very efficient shortcut explained below, it is suggested that *until* students understand the meaning of percent, they would do it on a calculator in a sequence similar to the method they would use without a calculator.

This problem might be used to introduce the worksheet.

In a special sale where all items were discounted 15%, a tennis racket regularly sold for $89.50. The sales tax is 5%. How much did the tennis racket cost during the sale?

Three methods of solving this problem follow.

1. Students must first understand that a percent such as 15% means 15/100 or .15. 15/100 of $89.50 is the amount of money which would be deducted from the list price. The student first puts the list price into memory (M+). Then he/she multiplies .15 × $89.50 and subtracts this value from the list price which is in memory (M−). R−CM shows that the cost of the item before tax is $76.075. Stores normally round the sales price before they compute the tax. Therefore students put the new value, $76.08, into memory. To determine the tax, they multiply $76.08 by .05. This is rounded up (as taxes usually are) and added to the number in memory (M+). R−CM returns the cost of the item to the buyer – $79.89. A good topic for discussion is the ultimate effect of the rounding on the final cost to the consumer.

2. A deduction of 15% from the cost of an item means the customer pays 85% of the price. Therefore, the cost of the tennis racket can be determine by finding 85% of $89.50 rather than by finding 15% and subtracting. Similarly, the price including the tax can be found by finding 105% (1.05) of $76.08.

3. The calculator provides a third, and possibly simpler, method of finding the cost of an item at 15% discount. We would suggest that this method be taught only to those students who really understand percent, because to some students, this method may seem to be a *magic* discount rather than a logical process. To find the cost of the tennis racket at 15%, put $89.50 into the calculator. Then press the following sequence of keys: $\boxed{-}$ 15 $\boxed{\%}$. Your screen should read 76.075. To add in the tax, put $76.08 into the calculator. Then press $\boxed{+}$ 5 $\boxed{\%}$. This will give you 79.884 which rounds to $79.89. The student must be careful not to use the $=$ key when using this method. Results will occur for which students at this grade level may not be prepared.

SOLUTIONS:

1. No. He is short $.71.

2. $25.05

3. She can buy anything except the toaster if she buys one item only. She can buy the blender and the electric knife, or the can opener and the electric knife. This last combination will leave her $1.17 in change.

4. Yes. He earned $29.55 and will have $.06 left over.

5. A can opener

6. a. $468.64

   b. $477.91 (36% is the average of all the different discounts.)

WHO LIVES WHERE?

According to the 1980 U.S. Census*, approximately 226,543,000 people lived in the United States.

The Census figures also showed which five states were most heavily populated and which five were least populated.

Most Populated States**		Least Populated States**	
California	23,668,000	Alaska	402,000
New York	17,558,000	Wyoming	470,000
Texas	14,226,000	Vermont	511,000
Pennsylvania	11,865,000	Delaware	594,000
Illinois	11,427,000	North Dakota	653,000

In the problems below, round your answers to the nearest tenth of a percent, if necessary.

1. Which state has about 10% of the U.S. population?

_____

2. Which state or states has less than 0.2% of the U.S. population?

_____

3. What percent of the U.S. population is located in the five most populated states?

_____

4. What percent of the U.S. population is located in the five least populated states?

_____

5. Is your state on the list? If yes, what percent of the U.S. population lives in your state? If no, predict the largest and smallest percent of the U.S. population your state could be.

_____

6. California became a state in 1850. At that time the U.S. Census showed about 93,000 people lived there. By what percent has the population of California increased?

_____

7. In 1970 New York had about 18,241,000 people. As you can see, it has lost some people. By what percent has its population decreased?

_____

*Source: **The World Almanac and Book of Facts, 1989**

** Data is rounded to nearest thousand.

WHO LIVES WHERE?

OBJECTIVE: The student will work with percent.

COMMENTS: Before handing out this worksheet, you might ask the children to guess which state has the largest population. Students should be asked to write the problems in the form of a ratio to emphasize that percent means *out of 100*. It is also important to discuss the need to round the larger numbers to thousands to use on the calculator. We recommend that the students attempt problems #1 and #2 by inspection and they check their guess using the calculator.

There are various ways to solve these problems as discussed in previous activities. Your students should 1) choose the methods most comfortable for them. We hope in follow-up discussions students will learn there is a symbolic way to represent these problems and 2) how to deal with a percent as large as the solution to problem #6.

The percent key on the calculator may be used for the first two problems.

SOLUTIONS:

1. California $\dfrac{10}{100} = \dfrac{x}{226,543,000}$

(Press 226543 $\times$ 10% on the calculator; then move the decimal point three places to the right.)

2. Alaska $\dfrac{2}{1000} = \dfrac{x}{226,543,000}$

3. 34.8% $\dfrac{x}{100} = \dfrac{78,746,000}{226,543,000}$

4. 1.2% $\dfrac{x}{100} = \dfrac{2,631,000}{226,543,000}$

5. Answers will vary.   Largest is a maximum of 5.0(4)%.
Smallest is a minimum of 0.3%.

6. 25349% increase $\dfrac{x}{100} = \dfrac{23,668,000 - 93,000}{93,000}$

7. 3.7% $\dfrac{x}{100} = \dfrac{18,241,000 - 17,558,000}{18,241,000}$

ADDITIONAL ACTIVITY:

If the student's state is not on the list, you might want to ask him/her to use an almanac to find the necessary data to answer problem #5. Also students could pose their own original questions about additional data from the almanac.

PARENTHESES

Complete each equation by putting in the appropriate operations and parentheses where needed.

Example: 48 _____ 36 _____ 12 = 24

Solution: $48 - 36 + 12 = 24$ or $48 - (36 - 12) = 24$

Both of these solutions can be obtained on the calculator. The first one is very straightforward. The second requires the use of the memory keys. Try it to be sure you understand how to use the memory.

48 [M+] 36 [−] 12 [M −] [RCM]

Your display should read 24.

Some of the problems below may have more than one solution. Sometimes you can solve a problem easily by hand, but the method will not produce an exact answer on the calculator.

Example: 20 _____ 15 _____ 9 = 12

You can write $20 \div 15 \times 9 = \dfrac{20}{15} \times 9 = \dfrac{4}{3} \times 9 = 12$

When you put this sequence into the calculator, you will get the answer 11.999999. Check this on your own calculator. (If you happen to own a scientific calculator, try it on that one. There is a difference.)

1. 48 _____ 36 _____ 12 = 72

2. 48 _____ 36 _____ 12 = 51

3. 48 _____ 36 _____ 12 = 1,716

4. 48 _____ 36 _____ 12 = 45

5. 48 _____ 36 _____ 12 = 1

6. 48 _____ 36 _____ 12 = 2

7. 48 _____ 36 _____ 12 = 2,304

8. 48 _____ 36 _____ 12 = 16

9. 48 _____ 36 _____ 12 = 7

10. 48 _____ 36 _____ 12 = 480

How many other equations can you write using 48, 36, and 12 in this order which will produce *different* whole number solutions? (There are at least 5 more.)

PARENTHESES

OBJECTIVE: The student should use rules of order and appropriately place parentheses to create true number sentences.

COMMENTS: The rules of order should be reviewed. The significant parts of the rules of order for this worksheet are:

- Working from left to right, complete all work within parentheses first.
- Working again from left to right, complete all multiplication and division problems *in the order in which they occur.*
- Working again from left to right, complete all addition and subtraction problems *in the order in which they occur.*

Here are some samples which can be used for overhead or chalkboard use to illustrate rules of order:

$12 + 6 - 4 + 2 = $ ____ (16) $20 \div 5 \times 4 \div 2 = $ ____ (8)

$12 + 6 - (4 + 2) = $ ____ (12) $20 \div (5 \times 4) \div 2 = $ ____ (½)

$12 + (6 - 4) + 2 = $ ____ (16) $20 \div (5 \times 4 \div 2) = $ ____ (2)

$12 - (6 - 4) - 2 = $ ____ (8)

$12 - 6 - (4 - 2) = $ ____ (4)

Here are some examples which can be used to illustrate the worksheet:

Possible solutions:

20 ____ 5 ____ 4 ____ $2 = 30$ $(20 - 5) \times 4 \div 2 = 30$

20 ____ 5 ____ 4 ____ $2 = 6$ $(20 \div 5) + 4 - 2 = 6$

20 ____ 5 ____ 4 ____ $2 = 38$ $20 + (5 + 4) \times 2 = 38$

SOLUTIONS:

There may be others which will work on the calculator to give an exact answer.

1. $48 + 36 - 12 = 72$

2. $48 + (36 \div 12) = 51$

3. $48 \times 36 - 12 = 1,716$

4. $48 - (36 \div 12) = 45$

5. $48 \div (36 + 12) = 1$ or $(48 - 36) \div 12 = 1$

6. $48 \div (36 - 12) = 2$

7. $48 \times (36 + 12) = 2,304$

8. $48 \div (36 \div 12) = 16$

9. $(48 + 36) \div 12 = 7$

10. $48 + (36 \times 12) = 480$

Some other whole number solutions which may be obtained by operating on this set of numbers in this exact order are 0; 96; 144; 1,152; and 20,736. Perhaps your students will find a few more.

ADDITIONAL ACTIVITY:

Ask students to create different problems with other sets of numbers and give them to classmates to try.

A NUMBER IS MISSING!

Fill each box with a whole number so that the sentence will be true, according to the rules of order of operations.

> **Example:** $16 - \boxed{} \times 3 + 4 = 8$
>
> **Solution:** $16 - 4 \times 3 + 4 = 8$

1. $24 \div \boxed{} \times 12 = 18$

2. $84 - 2 \times \boxed{} + 2 = 10$

3. $150 \div (3 + \boxed{}) - 8 \times \boxed{} = 1$ (same number in each box)

4. $38 < 459 \div \boxed{} + 12 < 40$

5. $182 < \boxed{} \times 54 \div 6 < 190$

6. $0 < (\boxed{} \div 2) - 14 < 1$

7. $37 < 425 - \boxed{} \times 43 < 40$

8. $1000 \div \boxed{} \times 2 + 17 = 142$

A NUMBER IS MISSING!

OBJECTIVE: The student will solve equations which require an understanding of rules of order. He/she will also use educated guessing as a problem-solving strategy.

COMMENTS: In general, students will solve these equations using trial and error methods. It is to be hoped that they will select two or three numbers, try them in the equations, notice what effects their choices have, and alter them to come closer and closer to the desired result.

The worksheet includes one example, but another possible example which might be used for classroom demonstration and discussion is the following:

$$80 + 36 \div \boxed{} \times 4 = 96$$

Students should first be given the opportunity to suggest solutions. Those who understand order of operations may come up with the correct answer which is 9.

Those who have difficulty may first need to be reminded of the rules for order of operations. Once it is clear that they should start with the first division, they can test some numbers. For example, if the first guess is 4, the answer would be 116, clearly too large. Then they might try 6, which would give the answer 104, which is still too large but is clearly moving in the right direction. When they now try the next larger divisor of 36, 9, they will find it is the correct solution.

Another way to help direct student thinking would be to ask questions like:

> What number added to 80 gives you 96? (16)
> What number times 4 will give you 16? (4)
> By what number can you divide 36 to get 4? (9)

SOLUTIONS:

1. 16

2. 38

3. 3

4. 17

5. 21

6. 29

7. 9

8. 16

FACTOR HUNT

Here are some challenging problems on which you can use your knowledge of factors. It will help to keep a pencil and paper handy to keep a record of the numbers you've already tried.

1. What number less than 200 has the largest number of whole number factors?

2. What numbers less than 300 have exactly 3 factors, including the number itself and 1?

3. What is the smallest number which has exactly 7 factors?

4. The following numbers have only one common factor, other than 1. What is it?

    22,098   16,891   37,465

5. 7 is not a factor of 4. It is not a factor of 44 or 444. Is there a number made up of 4's of which 7 is a factor?

What can you say about any other numbers made up of 4's of which 7 is a factor?

Can you find some types of numbers which can never be a factor of numbers whose digits are all 4's?

6. What is the greatest common factor (GCF) of these three numbers?

    19,019   101,101   510,510

7. I'm thinking of two numbers whose greatest common factor (GCF) is 14 and whose least common multiple (LCM) is 210. The numbers are not 14 and 210. What could these numbers be?

8. What is the smallest number for which all the numbers 2 to 15 are factors?

FACTOR HUNT

OBJECTIVE: The student will use his/her knowledge of factors and multiples to solve several challenging problems.

COMMENTS: Students who understand how to use prime numbers to find factors of a number will have little difficulty solving the first three problems. Problem #1, which is looking for the largest number of factors of a number less than 200, can most easily be solved if the student realizes that the solution must be factorable by a large number of prime numbers. The prime factorization of 144 contains 7 numbers. $2^4 \times 3^3$ = 144. When all combinations of these 7 numbers are found, there will be 13 whole number factors. When 1 and 144 are included, there is a total of 15 whole number factors.

In problem #2, students should be aware that only squares of prime numbers have exactly three factors.

In problem #3, those who realize that only square numbers have an odd number of whole number factors will be well on their way to finding the solution.

Knowledge of divisibility rules and prime numbers are needed to solve problem #4. Once students find the prime factors of one number, they should be able to test the other two numbers readily. This is also true of problem #6.

Problem #5 has the potential for generating some interesting class discussions. Students should be able to discover that multiples of 5 cannot be factors of a number whose digits are all 4's. Also, any multiple of 8 cannot be a factor. All efforts to divide any number of 4's by 8 will always lead to a remainder of 4. Teachers might also ask students to check other numbers with the same digits, 9's or 1's or 2's, for example. What divisors will work? Which will not work?

The product of any pair of numbers will always be the same as the product of the LCM and GCF of the two numbers. Students should be able to solve problem #7 by finding the prime factors of 14 and 210 and then finding which combination of these primes will produce the same LCM and GCF.

To solve problem #8, students must recognize that they need to find the smallest number of prime factors which will produce all of the numbers through 15. This set of primes is $2 \times 2 \times 2 \times 3 \times 3 \times 5 \times 7 \times 11 \times 13$.

SOLUTIONS:

1. 144 (It has 15 factors – 1, 2, 3, 4, 6, 8, 9, 12, 16, 18, 24, 36, 48, 72, 144)

2. 4, 9, 25, 49, 121, 169, 289 (the squares of prime numbers)

3. 64 (1, 2, 4, 8, 16, 32, 64)

4. 127

5. Yes, 444,444.
The number of 4's in the number must be a multiple of 6.
Numbers which are multiples of 5 and 8.

6. 1,001 (7 x 11 x 13)

7. 42 and 70

8. 360,360

SQUARE FACTS

1. Write down the squares of all numbers from 1 to 25.

$1^2 =$ _____ $10^2 =$ _____ $18^2 =$ _____

$2^2 =$ _____ $11^2 =$ _____ $19^2 =$ _____

$3^2 =$ _____ $12^2 =$ _____ $20^2 =$ _____

$4^2 =$ _____ $13^2 =$ _____ $21^2 =$ _____

$5^2 =$ _____ $14^2 =$ _____ $22^2 =$ _____

$6^2 =$ _____ $15^2 =$ _____ $23^2 =$ _____

$7^2 =$ _____ $16^2 =$ _____ $24^2 =$ _____

$8^2 =$ _____ $17^2 =$ _____ $25^2 =$ _____

$9^2 =$ _____

Write down all the digits which appear in the ones place.

_____

Which digits are missing?

_____

Can you think of any squares of whole numbers which have these missing digits in the ones place?

_____

2. Study each of these numbers. Check off those which could not possibly be squares of whole numbers. Then use your calculator to find out which of the others are square numbers. Write down the whole number for which each is a square.

Example: 256 ___16^2___

3,364 _____   2,187 _____   58,081 _____   41,681 _____

9,918 _____   3,249 _____   15,652 _____   15,625 _____

8,881 _____   1,444 _____   119,716 _____   217,144 _____

3. Look at the numbers 12 and 21. The digits in the first are in the reverse order in the second. Now look at their squares: 144 (12^2) and 441 (21^2). The digits of the squares are also in reverse order. Find as many pairs of three-digit numbers as you can where the digits in each are in reverse order *and* the digits in their squares are also in reverse order. What are these numbers? How many pairs did you find?

_____

_____

SQUARE FACTS

OBJECTIVE: The student will learn about some characteristics of square numbers.

COMMENTS: This worksheet should provide excellent material for group discussion. The problems are probably self-explanatory as long as students understand the meaning of the exponent. If this is not familiar to them, it should only take a few minutes to show that 3^2 means 3×3 and 4^2 means 4×4.

One discovery students should make is that only certain digits will appear in the ones place of any square number. If they see a number which ends in 3, for example, they should know immediately that this number cannot be a square number. They should be able to check off those numbers which cannot be square numbers without any difficulty. If the teacher is interested in teaching the square root symbol and function, the student can determine whether a number is a square by using this function on the calculator. This is not necessary, however. Students should be able to look at a number like 841, decide that the digit in the ones place can only be a 1 or a 9, and then check only numbers between 20 and 30, since 841 lies between 20^2 (400) and 30^2 (900). The two numbers to be checked are 21 and 29. $21^2 = 441$ and $29^2 = 841$.

In problem #2, students should recognize that the only numbers which will work will have some combinations of the digits 0, 1, 2, and 3.

SOLUTIONS:

1.

1	100	324
4	121	361
9	144	400
16	169	441
25	196	484
36	225	529
49	256	576
64	289	625
81		

0, 1, 4, 5, 6, 9

2, 3, 7, 8

No

2.

58^2	✔	241^2	no
✔	57^2	✔	125^2
no	38^2	346^2	no

3.   102 (10,404) and 201 (40,401)
103 (10,609) and 301 (90,601)
112 (12,544) and 211 (44,521)
113 (12,769) and 311 (96,721)
122 (14,884) and 221 (48,841)

ADDITIONAL ACTIVITY:

Ask students to investigate the cubes of the numbers from 1 to 25. Are there any missing digits as there are with the squares of these same numbers?

THE POWERS THAT BE

> In the expression 2^7, 2 is called the *base*. 7 is called the *exponent*.
>
> 2^7 means $2 \times 2 \times 2 \times 2 \times 2 \times 2 \times 2$ which means 2^7 equals 128.
>
> 128 is called a *power* of 2.

In all problems below, you may select only a number between and including 2 and 10, as the base, and a number between and including 2 and 6, as the exponent.

1. $4^2 = 2^4$ (16 = 16). These are equivalent expressions. Using the numbers given above, how many other equivalent expressions can you find? Record what you find. These expressions may help you in problem #2 below.

_____

_____

_____

_____

2. In each number sentence below, fill in the blanks with expressions using exponents and bases which will make the sentence true.

   a. _____ + _____ = 5^2

   b. _____ + _____ = 10^2

   c. _____ − _____ = 10^2

   d. _____ + _____ + _____ = 3^3

   e. _____ − _____ = $2^2 + 6^2$

   f. _____ + _____ + _____ = 10^3

   g. _____ − _____ − _____ = 22^2

   h. _____ − _____ − _____ = 12^2

THE POWERS THAT BE

OBJECTIVE: The student will work with different powers of numbers and will discover some relationships between powers of different numbers.

COMMENTS: This worksheet will give the students the opportunity to experiment with powers of the numbers between and including 2 and 10. Students should quickly discover, if they are not already aware, that 2, 4, and 8 have a special relationship, since both 4 and 8 are powers of 2. It should become clear to them, as they study the example, that since $2^4 = 2 \times 2 \times 2 \times 2$ and $4^2 = 4 \times 4$, these expressions are equivalent. Similarly, $4 \times 4 \times 4 = 2 \times 2 \times 2 \times 2 \times 2 \times 2 = 8 \times 8$ or $4^3 = 2^6 = 8^2$. All of these expressions are related because they involve powers of 2.

They should become aware also of the special relationship between 3 and 9. Since this worksheet limits students to numbers which are no larger than 10, there are no other worksheet examples to probe this pattern further. However, the calculator is such an excellent means of exploring powers that the teacher might suggest students look for other such patterns as $5^4 = 25^2$ and $4^{12} = 16^6$.

Without using the calculator, can they determine the following missing exponents?

        a. $9^6 = 81^?$ (3)   **b.** $2^{21} = 8^?$ (7)

To find solutions for the second part of the worksheet, students should use the information developed in the first part and they will also need to use their estimation skills.

SOLUTIONS:

1. $2^6 = 4^3$ $3^4 = 9^2$ $4^3 = 8^2$

    $2^6 = 8^2$ $3^6 = 9^3$ $4^6 = 8^4$

2. Answers may vary. Alternative solutions may be obtained when equivalent expressions are substituted.

    a. $3^2 + 4^2 = 5^2$ $9 + 16 = 25$

    b. $6^2 + 8^2 = 10^2$ $36 + 64 = 100$

    c. $5^3 - 5^2 = 10^2$ $125 - 25 = 100$

    d. $3^2 + 3^2 + 3^2 = 3^3$ $9 + 9 + 9 = 27$

    e. $4^4 - 6^3 = 2^2 + 6^2$ $256 - 216 = 4 + 36 = 40$

    f. $5^4 + 7^3 + 2^5 = 10^3$ $625 + 343 + 32 = 1000$

    g. $5^4 - 5^3 - 4^2 = 22^2$ $625 - 125 - 16 = 484$

    h. $8^3 - 7^3 - 5^2 = 12^2$ $512 - 343 - 25 = 144$

LARGE NUMBER SHORTHAND

Try these on your calculator. See if you can find the exponent or base number to make the equation true.

1. $6^? = 279,936$ _____

2. $?^8 + 10^4 = 10,256$ _____

3. $2^{13} - ?^6 = 4,096$ _____

4. $10^3 + 12^2 + 4^? + 6^4 + 5^2 = 6,561$ _____

5. $2^? + 4^5 + 10^3 + 5^2 - 1^{11} = 4,096$ _____

6. $?^5 + 5^2 - 6^4 - 3^3 = 31,470$ _____

7. $6^5 - 4^6 - ?^5 - 8^3 - 3^3 - 4^2 = 0$ _____

LARGE NUMBER SHORTHAND

OBJECTIVE: The student will understand the meaning of an exponent and use exponents within a mathematical sentence.

COMMENTS: Discuss how the constant function of multiplication can be used to find higher powers of numbers. For example, to find 2^5, press $2 \times = = = =$. Emphasize that exponents are done before the processes of multiplication, division, addition, and subtraction according to order of operations.

The addition and subtraction operations with exponents provide an opportunity to use the memory key to store information while doing another operation.

SOLUTIONS:

1. 7

2. 2

3. 4

4. 6

5. 11

6. 8

7. 5

WHAT ARE MY NUMBERS?

1. I'm thinking of a number...

If you put my number in place of the question marks in the number sentence below, the sentence will be true.

$3^? + ?^3 - 2^? - ?^4 = 1$

What is my number? _____

2. I'm thinking of two numbers...

The difference of the squares of my two numbers is a cube.
The difference of the cubes of my two numbers is a square.

What are the smallest two numbers I could be thinking of? _____

3. I'm thinking of a number...

The square of the cube root of my number is a whole number.
The cube of the square root of my number is a whole number.
My number is not 64.

What is the next smallest number my number could be? _____

4. I'm thinking of a number...

If you add the cubes of each digit in the number 407 ($4^3 + 0^3 + 7^3$), the sum is 407.
The cubes of the digits of my number will also add up to my number.
My number is between 350 and 400.

What is my number? _____

5. I'm thinking of a number...

My number is the number of days it will take me to save $100,000.00 if I start saving a penny today, two cents tomorrow, four cents the next day, and every day from then on, I continue to double the amount.

What is my number? _____

WHAT ARE MY NUMBERS?

OBJECTIVE: The student will solve a series of challenges which use powers of numbers.

COMMENTS: In order to solve these problems, the student should be familiar with powers of numbers and also with the concepts of square roots and cube roots. Problems #1 and #2 will be trial and error for most students, but they might organize their work by starting with 2 and then using consecutive numbers until they reach the solution. The first problem provides an excellent opportunity to use the memory key on the calculator.

Problem #3 requires an understanding of the meaning of square root and cube root. Probably the best way to do this is to find the cube of each number in order starting with 2. Then check the cube using the square root key to find out if it is also a square number.

To solve problem #4 it will be helpful to find the cubes of all the digits first. Once the students have done this, they can eliminate the cubes of 8 and 9 because they are too large. They should also note that one of the digits must be 3 since the number lies between 350 and 400. Using their list of cubes, they should then be able to eliminate and narrow their results until they reach the answer(s).

Problem #5 is a familiar problem phrased slightly differently. The class will probably have a variety of techniques for solving this and an interesting discussion should ensue. Perhaps the easiest way to solve it is to count the number of times 2 is used as a product to reach a value greater than $100,000.00.

$$2^{24} = \$167,772.16$$

The actual amount saved at the end of the 24th day will be $167,772.15.

SOLUTIONS:

1. 7 $(3^7 + 7^3 - 2^7 - 7^4 = 1)$

2. 10 and 6

3. 729

4. 370 or 371

5. 24 days

ESTIMATE! CALCULATE! EVALUATE! © 1990 CUISENAIRE COMPANY OF AMERICA, INC.

BE AN EFFECTIVE DETECTIVE

For each set of clues, the box [ ] represents the same number. Find the mystery number.

1. a. I am a prime number between 400 and 500.

d. [ ] $> 6,570 \div 15$

b. [ ] $> 45,271 - 44,863$

e. Two of my digits are the same.

c. [ ] $< 219 + 268$

f. [ ] $< 4,917 \div 11$

Who am I? _____

2. a. I am an even number.

e. [ ] $< 5^4$

b. I am a multiple of 3.

f. None of my digits is the same.

c. [ ] $> 23^2$

g. My tens digit is the largest.

d. Five is one of my factors.

Who am I? _____

3. a. I am a prime number.

d. [ ] $< 7,584,693 - 7,583,942$

b. [ ] $> 62,389 \div 89$

e. I am halfway between two other primes.

c. None of my digits is a placeholder.

Who am I? _____

4. a. I am a four-digit number.

d. None of my digits are the same.

b. I am a multiple of 25.

e. None of my digits is a square number.

c. I have an odd number of factors.

Who am I? _____

BE AN EFFECTIVE DETECTIVE

OBJECTIVE: The student will solve number theory problems using the strategy of logical elimination.

COMMENTS: This activity uses a variety of skills involving the four basic processes with whole numbers. Inequalities, factors, primes, multiples, exponents, and place value are also components of this exercise.

Some students will probably need a review of some of these concepts before trying this worksheet. For example, if they are locating the prime numbers between 100 and 150, review of rules of divisibility will be useful. Students should then be able to determine by inspection which numbers are divisible by 2, 3, and 5. At this point, the calculator becomes a valuable tool to discover higher primes.

It would also be advantageous to have students look at the numbers one through twenty-five to discover what type of numbers provide an odd number of factors.

Some of the students should only try problem #1 while others may wish to attempt all four.

SOLUTIONS:

1. 443

2. 570

3. 733

4. 3,025

ADDITIONAL ACTIVITY:

Students can write problems of their own for either the class or another classmate. This can prove to be a very valuable learning experience. You might stress the importance of allowing for a range of answers so that the actual number cannot be determined until the final clue.

BE A VERY EFFECTIVE DETECTIVE

For each set of clues, the box [ ] represents the same number. Find the mystery number.

1. a. I am an odd number.

   b. All of my digits are the same.

   c. [ ] $< 10^3$

   d. I am not divisible by 9.

   e. I have more than one digit.

   f. I am not a multiple of 11 or 5.

   g. My digit is a prime number.

Who am I? _____

2. a. I am a composite number.

   b. [ ] $< 27^2$

   c. [ ] $> \sqrt{436,921}$

   d. I am not divisible by 2, 3, 5, or 7.

   e. The sum of my digits is 14.

Who am I? _____

3. a. I am a four-digit even number.

   b. My digits are in descending order (not necessarily consecutive).

   c. My thousands digit is twice as large as my tens digit.

   d. My only prime factors are 2, 3, and 5.

   e. [ ] $> 256^2 - 246^2$

Who am I? _____

4. a. I am a four-digit palindromic number.

   b. I am odd.

   c. My smaller prime factor is a factor of all four-digit palindromic numbers.

   d. Both of my prime factors are also palindromic numbers.

   e. Two of my digits are consecutive whole numbers.

   f. One of my prime factors is greater than 150 but less than 350.

Who am I? _____

BE A VERY EFFECTIVE DETECTIVE

OBJECTIVE: The student will solve number theory problems using the strategy of logical elimination.

COMMENTS: This activity requires some of the same skills needed in Activity 34 **Be An Effective Detective**. But, in addition, knowledge of palindromic numbers is needed. Palindromic numbers are numbers that read the same forwards and backwards such as 919 or 4,224.

SOLUTIONS:

1. 777

2. 671

3. 8,640

4. 3,443

ADDITIONAL ACTIVITY:

See Activity 34 **Be An Effective Detective** on page 75.

CHECKING YOUR CHECKS

Evan is trying to balance his bank account. Use the checkbook form on page 88 to help him validate his monthly statement.

1. Balance on hand as of 12/1 was $5,267.38.

2. Wrote check on 12/2 for $342.96 to Dr. Fixit.

3. Deposited on December 4, $617.52 received from sales.

4. Wrote check dated 12/5 for $538.39 to Ace Electronics for new television.

5. On December 6 wrote check for $47.94 to L. B. Goods for packaging material.

6. Wrote check for $1,205.64 on the following day to Big Deal Appliance for new computer.

7. Bank paid $27.89 interest on bank account as of 12/10.

8. On December 15 deposited $2,567.57 received from sales.

9. Salary check deposited on 12/16 for $753.49.

10. On 12/19 bank charged ten dollars for lost check.

11. Wrote check for $319.93 on the twenty-first to C. Software Co. for new programs.

12. Wrote check for $2,036.93 two days later to South Co. for printers.

13. The following day wrote check for $1,577.25 to A. Gyp.

14. Received a birthday check for $25.00 and cashed it.

15. Safe deposit box rental fee of $35.00 paid by check on 12/28.

16. Monthly service charge on 12/30 for $7.50.

Balance as of 12/31: _____

CHECKING YOUR CHECKS

OBJECTIVE: The student will organize data in a chart, and add and subtract mixed decimal fractions.

COMMENTS: It is necessary that students understand the importance of organizing data for easy reference. Discuss the form of a check book and how entries are made.

SOLUTIONS:

PLEASE BE SURE TO DEDUCT ANY PER CHECK CHARGES OR SERVICE CHARGES THAT MAY APPLY TO YOUR ACCOUNT.

CHECK NO.	DATE	CHECKS ISSUED TO OR DESCRIPTION OF DEPOSIT	(−) AMOUNT OF CHECK		T	(−) CHECK FEE (IF ANY)	(+) AMOUNT OF DEPOSIT	BALANCE	
								5,267	38
101	12/2	Dr. Fixit	342	96				4,924	42
	12/4	Money from sales					617 52	5,541	94
102	12/5	Ace Electronics	538	39				5,003	55
103	12/6	L. B. Goods	47	94				4,955	61
104	12/7	Big Deal Appliance	1,205	64				3,749	97
	12/10	Interest					27 89	3,777	86
	12/15	Money from sales					2,567 57	6,345	43
	12/16	Salary					753 49	7,098	92
	12/19	Lost check fee				10.00		7,088	92
105	12/21	C. Software Co.	319	93				6,768	99
106	12/23	South Co.	2,036	93				4,732	06
107	12/24	A. Gyp	1,577	25				3,154	81
108	12/28	Safe deposit rental	35	00				3,119	81
109	12/30	Service charge	7	50				3,112	31

REMEMBER TO RECORD AUTOMATIC PAYMENTS / DEPOSITS ON DATE AUTHORIZED.

WHO FITS THE PROFILE?

Three candidates are running for President of the United States.

- • Candidate A was born in Virginia, is 44 years old, married with 2 children, a Democrat.

- • Candidate B was born in Ohio, is 56 years old, married with 3 children, a Republican.

- • Candidate C was born in Massachusetts, is 64 years old, married with 6 children, a Republican.

Below is a chart giving information about the 41 Presidents of the United States.

Use this information to answer the questions on page 83.

Name	Birth-place	Age at Inauguration #	Party*	Name	Birth-place	Age at Inauguration #	Party*
Washington	VA	57	F	Cleveland	NJ	47	D
Adams, J.	MA	61	F	Harrison, B.	OH	55	R
Jefferson	VA	57	DR	Cleveland	NJ	55	D
Madison	VA	57	DR	McKinley	OH	54	R
Monroe	VA	58	DR	Roosevelt, T.	NY	42	R
Adams, J.Q.	MA	57	DR	Taft	OH	51	R
Jackson	SC	61	D	Wilson	VA	56	D
Van Buren	NY	54	D	Harding	OH	55	R
Harrison, W.	VA	68	W	Coolidge	VT	51	R
Tyler	VA	51	W	Hoover	IA	54	R
Polk	NC	49	D	Roosevelt, F.	NY	51	D
Taylor	VA	64	W	Truman	MO	60	D
Fillmore	NY	50	W	Eisenhower	TX	62	R
Pierce	NH	48	D	Kennedy	MA	43	D
Buchanan	PA	65	D	Johnson, L.	TX	55	D
Lincoln	KY	52	R	Nixon	CA	56	R
Johnson, A.	NC	56	R?**	Ford	NE	61	R
Grant	OH	46	R	Carter	GA	52	D
Hayes	OH	54	R	Reagan	IL	69	R
Garfield	OH	49	R	Bush	MA	64	R
Arthur	VT	51	R				

*Party abbreviations: F – Federalist   DR – Democrat-Republican   D – Democrat
                         W – Whig   R – Republican

#Age at time of inauguration for first term, if the same person served consecutive terms. Grover Cleveland is included twice because he served four years, was out of office for a term and then was elected again.

**Andrew Johnson was a Democrat who was nominated as Vice President by the Republicans and took office when Lincoln was assassinated.

WHO FITS THE PROFILE?

OBJECTIVE: The student will make a prediction using information gained by calculating the means, medians, and modes of given sets of data.

COMMENTS: The student is asked, in this activity, to use information about former Presidents of the United States to select from a list of three candidates the one who would be most likely to fit the *Presidential Profile*. If these statistics alone were used, Candidate B would be elected. It is to be hoped, however, that students will realize that any prediction here is being made just for fun, and that all the elements that go into the choosing of a president could not possibly be listed here.

In fact, while Candidate B fits our profile almost perfectly, it is Candidate C who most closely resembles President George Bush. Nevertheless, students might be aware that there are some clues to be found in these figures. They would be justified in concluding that a man under 40 or over 70 would be unlikely to be elected. Also, it seems clear that the majority of our Presidents come from the larger, more heavily populated states.

Students are asked to use three measures of central tendency, the mean, the median, and the mode. The mean, of course, is the average, found by totaling all the values and dividing by the number of values. The median is the middle score, determined by ordering all the values and finding the middle value. The mode is the value which occurs most frequently. In the ideal statistical world, the mean, median, and mode would all be the same because the data would be represented by a normal bell-shaped curve. However, in this set of data, as in most data with which the students will work, there are differences which should be noted. As can be seen in the solutions, the mean and median are the best measures of central tendency for this data.

Students should be able to find from the graph the number of presidents who had no children, one child, two children, etc. They should then be able to determine how many children in all were born to all U.S. Presidents (143). They also should learn from the graph that there are only 40 people on the Presidential listing. A note at the bottom of the chart on the first page indicates that Ronald Reagan is considered our 40th President and George Bush our 41st President because Grover Cleveland was elected to two terms which were separated by four years. He is therefore considered to have been our 22nd and our 24th President. The graph of Presidents' children shows clearly that the modal value (2) is not necessarily the best measure of central tendency. And in fact, the mean is somewhat distorted because John Tyler had 15 children and Rutherford B. Hayes had 10 children. The median is, therefore, probably the best indication of the number of children born to Presidents (3).

WHO FITS THE PROFILE?

1. What is the average age of a President of the United States at the time of inauguration?

_____

The median age?_____ The modal age?_____

Which figure, the mean, median, or mode, do you think most accurately represents the typical age of a President at inauguration?

_____

2. In what state or states were the greatest number of Presidents born?

_____

3. Which party was favored by the voters in the majority of the Presidential elections?

_____

4. Use the graph below to answer these questions.

What is the average number of children born to presidents of the United States?

_____

The median? _____ The mode? _____

Which do you think is the best indication of the number of children born to U.S. Presidents – the mean, median, or mode?

_____

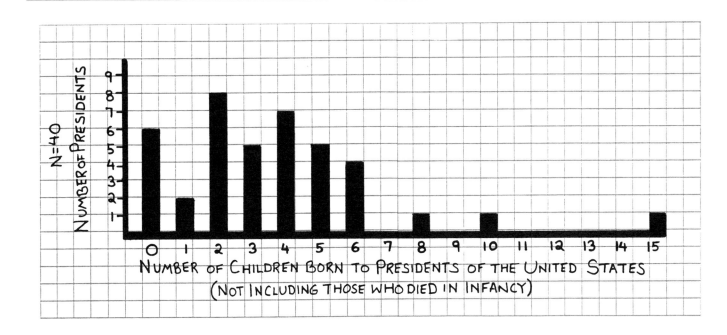

According to the data you just looked at, who is most likely to get elected, in your judgment – Candidate A, Candidate B, or Candidate C? Why?

_____

_____

WHO FITS THE PROFILE?

SOLUTIONS:

1. Mean (average)   55.07
   Median   55
   Mode   51

Both the mean and the median indicate that 55 is the *average* age of men elected to the Presidency.

2. 8 Presidents were born in Virginia, 7 in Ohio, 4 in Massachusetts, and 4 in New York. It should be noted that 7 of the first 12 Presidents were born in Virginia, and that since that time, only one was born there. In recent years the Presidents have come from a much more varied group of states.

3. In the last 100 years (since 1889), 11 Presidents have been affiliated with the Republican Party, 7 with the Democratic Party.

4. Among them, the Presidents had 143 children. As was noted before, for this measure, there are only 40 people represented.

   Mean   3.575
   Median   3
   Mode   2

As was mentioned on page 82, the median, 3, is probably the best measure of central tendency using this data.

If students were to use this data alone, Candidate B, 56 years old, born in Ohio, a Republican, with 3 children, best fits the profile drawn by the data from past presidents. However, the students should note that President George Bush is described exactly by Candidate C.

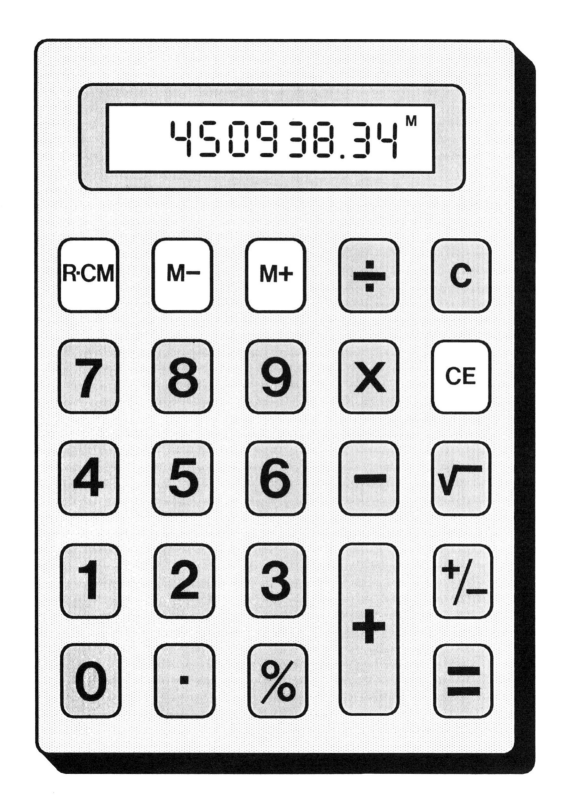

ZERO IN ON THE TARGET

SCORE SHEET

ROUND #	TARGET	MY NUMBER	MY TOTAL	OPPONENT TOTAL	DIFFERENCE BETWEEN EACH TOTAL & TARGET	
					MINE	OPPONENT
1	400					
2	850					
3	1000					
4	250					
5	750					
6	0					
				Totals		

ESTIMATE AND CALCULATE
SCORE SHEET

Your score for each round is the difference between your estimate and the actual solution. The final *lowest* total score wins.

GAME NUMBER _____

Round	Player #1	Player #2	Player #3
1	_____	_____	_____
2	_____	_____	_____
3	_____	_____	_____
4	_____	_____	_____
5	_____	_____	_____
6	_____	_____	_____
TOTALS:	_____	_____	_____

GAME NUMBER _____

Round	Player #1	Player #2	Player #3
1	_____	_____	_____
2	_____	_____	_____
3	_____	_____	_____
4	_____	_____	_____
5	_____	_____	_____
6	_____	_____	_____
TOTALS:	_____	_____	_____

CHECKING YOUR CHECKS

PLEASE BE SURE TO DEDUCT ANY PER CHECK CHARGES OR SERVICE CHARGES THAT MAY APPLY TO YOUR ACCOUNT.

CHECK NO.	DATE	CHECKS ISSUED TO OR DESCRIPTION OF DEPOSIT	(−) AMOUNT OF CHECK	✓ T	(−) CHECK FEE (IF ANY)	(+) AMOUNT OF DEPOSIT	BALANCE	

REMEMBER TO RECORD AUTOMATIC PAYMENTS / DEPOSITS ON DATE AUTHORIZED.